How to Burn a Goat

How to Burn a Goat

FARMING WITH THE PHILOSOPHERS

Scott H. Moore

BAYLOR UNIVERSITY PRESS

© 2019 by Baylor University Press
Waco, Texas 76798

All Rights Reserved. No part of this publication may be reproduced, stored in a retrieval system, or transmitted, in any form or by any means, electronic, mechanical, photocopying, recording, or otherwise, without the prior permission in writing of Baylor University Press.

Cover and interior design by Savanah N. Landerholm

First issued in paperback in 2023 under ISBN 978-1-4813-115-3

The Library of Congress has cataloged the hardcover as follows:

Library of Congress Cataloging-in-Publication Data

Names: Moore, Scott H., 1964- author.
Title: How to burn a goat : farming with the philosophers / Scott Moore.
Description: Waco : Baylor University Press, 2019. | Summary: "Literary and philosophical reflections combine with true-life farm anecdotes to offer commentary on seeking the good life in the modern age"-- Provided by publisher.
Identifiers: LCCN 2019015109 (print) | LCCN 2019980837 (ebook) | ISBN 9781481311526 (hardcover) | ISBN 9781481311557 (mobi) | ISBN 9781481311540 (epub) | ISBN 9781481311564 (pdf)
Subjects: LCSH: Christian philosophy. | Agriculture--Religious aspects--Christianity.
Classification: LCC BR100 .M677 2019 (print) | LCC BR100 (ebook) | DDC 191--dc23
LC record available at https://lccn.loc.gov/2019015109
LC ebook record available at https://lccn.loc.gov/2019980837

for Andrea

> ... We are more together
> than we know, how else could we keep on discovering
> we are more together than we thought?
> You are the known way leading always to the unknown,
> and you are the known place to which the unknown is always
> leading me back.
>
> —Wendell Berry, "The Country of Marriage"

CONTENTS

Preface	ix
A Burnt Offering	1
Buying Geese	3
Coming to Terms with Shit	5
Lambing	9
Guinea Fowl	11
Red in Tooth and Claw	15
Playing by Ear	17
HOMECOMING *and the Future of Higher Education*	21
Doing It and Getting It Done	29
The Fallacy of Acquisition	33
Mules	37
Septic Matters	41
Orphan Chicks	43
Silky Smooth's Big Adventure	47
To a Hare, from a Louse	55
FARMERS, CHRISTIANS, AND INTELLECTUALS *Cultivating Humility and Hope*	61
New Guineas	75

Skunks	77
Rattlesnakes	81
Dead Lambs	85
Alexander McCall Smith	89
How Many Chickens? How Many Eggs?	93
Ockham, Iris, and the Show Cattle	97
WENDELL, GENE, AND JOEL *On the Difficulties of Theology and Agriculture*	103
Do Sweat the Small Stuff	115
Not So Humble, but Near to the Ground	119
Saving Spiders	125
Snakes and Chicks	129
Tolstoy and Pahom	133
The Cow in the Parking Lot	135
BACK TO THE ROUGH GROUND *The Consolations of "Techne"*	139
Calves	155
E. B. White's Adventures in Contentment	159
Gussie, Lloyd, and Mocha	163
In Defense of Watching Grass Grow	167
Orchards	171
City of Sows	175
FARMING WITH THE PHILOSOPHERS *Work, Leisure, Wonder, and Gratitude*	179
Appendix: Iris Murdoch's Vexed Relationship with Christian Faith	193
Notes	203

PREFACE

"If you have a garden and a library, you have everything you need."

—Cicero

This great line from Cicero is certainly one of my favorite quotations. Unfortunately he was wrong. You will also need friends. And rain. And perhaps a chain saw.

In 2011 my wife and I bought a small ten-acre farm in Crawford, Texas. There was a cottage farmhouse (portions of which dated from the turn of the twentieth century) in desperate need of renovation, four barns in various states of disrepair, eleven bearing pecan trees, and one magnificent red oak more than half a century old. At that point in our lives, we were journeying through the dark night of the soul. We needed work, and the farm provided plenty of it for us.

The "garden and library" metaphor was appropriate for us in another way. Long before we bought the farm, we read farm books. We read all sorts: the how-to books, the memoirs, the nonfiction criticism, and the novels and stories of farm and country life. When faced with how to solve one particular problem on our farm, my father-in-law wryly replied, "Well, Scott can probably find a book about it." When people would ask us why we moved to the farm, one of my standard answers was that "you can only read Wendell Berry for so long before you finally give in."

Of course "giving in" takes different forms. For us it meant no longer pining away with the dream of "living in the country" but actually taking the plunge. For much of our married life, Andrea and I would say, "One day we're going to move to the country." She had grown up with a farm in her family, and it remained a place of retreat and blessing for all of us. We reared our five children mostly in the suburbs but always with a love of nature and creation and perhaps an overly sentimental view of what life on the farm would be like.

This book started with a few emails to a handful of friends telling stories of what was happening on the farm. Since we frequently didn't know what we were doing, the stories were sometimes rather ridiculous. What do you do with 150-pound dead goat? Who in their right mind would transport three large geese *inside* their car for more than a hundred miles? How do you catch an angry cow in a plowed field or a crowded parking lot? The best title for a book of these kinds of stories and essays had already been taken five hundred years ago. In 1509 Desiderius Erasmus wrote a little book entitled *In Praise of Folly*. It might have applied to us.

In Wendell Berry's *Imagination in Place*, he writes, "You can't learn to farm by reading a book. You can't lay out a fence line or shape a plowland or fell a tree or break a colt merely according to category; you are continually required to consider the distinct individuality of an animal or a tree, or the uniqueness of a place or situation."[1] Berry is bound to be correct about this, but he underestimates just how ignorant some of us are in these matters. We need language—both that printed in books and that spoken by neighbors—to give us the words to know how to name and understand what we are doing. Martin Heidegger famously said that "language is the house of Being." Without language we can't name (or know) what we experience.

For years I've had students say to me (about their term papers), "Dr. Moore, I know what I think, I just can't put it into words." And my stock response has always been, "Until you can put it into words, you don't yet 'know' what you think." I think that response is mostly true, but it's also misleading in some important ways. There are certain kinds of knowledge that defy our ability to "put it into words." Chief among these is the intellectual virtue of *techne*, the art and skill of learning how to create, repair, and engage the tools and objects of our lives. Attempting to understand the nature and limits of *techne* is one of the abiding themes of this book.

But in large measure, this book is an attempt to take my own advice, including discerning what can and cannot be adequately put into words. Telling farm stories and exploring the philosophical and cultural questions that inevitably arise amidst a life lived on the farm is one way of learning what we think and attempting to answer the age-old question, "How then shall we live?" The farm and the natural world is its own treasure house of Being that invites us into a rich journey of discovery and engagement.

Many of these reflections and essays have been circulated among friends, and I am deeply grateful for their suggestions and comments. Some friends have simply been a great encouragement to us on our journey. I am especially indebted to David and Lou Solomon (who came up with the title), Michael Beaty, Jeffrey Bilbro, Todd Buras, Darin Davis, Philip Donnelly, Michael and Alexandra Foley, Barry Harvey, Doug Henry, Tom Hibbs, Jenny and Eric Howell, David Lyle Jeffrey, John Nagy, Steve and Jane Nierman, John O'Callaghan, Mike Stegemoller, Cooper Thornton, Kay Toombs, Fr. Timothy Vaverek, Jeff Wallace, Roger Ward, and Ralph Wood.

So many people have been instrumental in helping us learn how to farm and how to appreciate life on the farm. Some of their stories and our debts to them appear in these pages. We are especially grateful to Art Hunter, the agriculture teacher and mentor at Crawford High School, without whom we would have accomplished very little on this journey. We are grateful to Betty and Buster Burleson, Lyndon and Tracy Love, Jacob Ray, Debbi Brown of "Unicorner Farm," Mackie and Norma Jean Bounds of "Swinging B Ranch," John and Heather Long, John and Sally Martin, the whole Dechaume family (especially Johnny and Barbara, Clinton, Jason, Eugene, and Patty), Charlie Kasparian, Janet Vance, Michael and Jessie Matsumoto, and Matt and Missy Tilghman.

This collection would not exist without Carey Newman and the staff (past and present) at Baylor University Press. Carey took an inchoate collection of farm stories and academic essays and brought them together into an (almost) coherent whole. I will always be grateful for his encouragement and his vision for this volume. Thanks also to Jordan Rowan Fannin, Emily Brower, Cade Jarrell, and Jenny Hunt.

Our debts to our family are our greatest. Andrea's parents, Ron and Ann Harrell, introduced me to the joys of the farm when I joined their family more than thirty years ago. "Grannie Annie" was the inspiration for so much of what our farm has become, and Ron ("Pappy") has patiently mentored us in so many of the construction and maintenance skills that are necessary to keep a farm running. Apart from their generosity, we would never have been able to begin this adventure. My own parents, Andy and Rachel Moore, have always been my models for what a life of hospitality and the love of learning look like. Growing up there was never a time when our home wasn't filled with students and guests who needed a home away from home. They taught me both the life of faith and the love of literature that infuse my life and this book. They introduced me to wonder, and it was they who taught me to see, like Wordsworth, "with an eye made quiet by the power / Of harmony, and the deep power of joy, / We see into the life of things."

Our five (now adult) children Emily Anne, Benjamin, Hannah, Samuel, and Andrew have given immeasurably to our farm experience. Each of them has contributed to the farm and our life here in their own ways. We have often reflected on how our children seem to replicate our lives in their own unique keys. In the early years, they spent countless afternoons, evenings, and weekends helping us renovate the Cottage and its barns and pens, and as the farm has grown and expanded, they have found their own special areas of interest in which to apply their unique talents and abilities to address our many needs. As their spouses have come into the family, they, too, have pitched in right along with all of us. We are so grateful for Shaun, Drew, and Phoebe. And of course the farm isn't all work. We have a fair amount of fun as well.

All of our friends know that Andrea is the real farmer around here. She's the one who tends to a thousand daily tasks: keeping the menagerie fed and watered, regularly mowing and mulching, and all the while making the farm a place of hospitality and hope for so many. She also manages the Airbnb out of the Cottage and is an extraordinary writer in her own right. She has edited and reedited these stories with me time and again. Almost every summer while I am gone for a month at a time teaching overseas, Andrea takes care of *all* the farming duties, doing battle with the varmints and the bugs and the weeds during the hottest and driest time of the year.

If these reflections ever give the impression that somehow she is only a helpmate and not at the center and the source of all that comes and goes on this farm, then my reporting will be false and unjust. Just as our evenings are often spent planning and discussing what's next on the farm, so our best days have been filled with those countless hours of learning how to build fences together, of helping young lambs come into the world and then shearing them the following spring, of planting and tending the gardens, of chasing, catching, and clipping the wings of chickens who terrorize those very gardens, and so very much more. Dreaming, planning, creating, working, and enjoying this farm with her is without a doubt the most pleasurable and fulfilling experience of my life.

It took a few years for us to come up with a name for our farm. Andrea and I wanted to find a name that would express both our gratitude and our hope for this small plot of land that we wanted to be a blessing to our family and community. Such a name should reflect both something of where we have come from and where we want to go. We wanted it to reflect the highest aspirations of our church, community, and university lives without seeming aloof or pretentious to our new neighbors. After much reflection we chose the name "Benedict Farms."

Why "Benedict Farms"? First, this farm is a blessing, a *benedictio*. As noted above we would not have been able to purchase it without the blessing that came from the estate of Andrea's mother, Ann Furr Harrell. Grannie Annie was and is a great blessing to us and to all who knew her. Moreover, the farm has been a blessing of renewal and restoration for us as we have given ourselves to the many and varied tasks of restoring the Cottage, barns, and fields of the farm. This hard work is helping us learn to live in a world touched by illness, loss, and recovery, and it's a blessing.

Second, Andrea and I both came independently to the recognition that we wanted the farm to reflect the virtues and practices of St. Benedict. In short it's our desire to make the farm a place of prayer, work, study, hospitality, and renewal. We do not yet know all the ways this farm will be used, but we know that we want to cultivate and to be formed by these practices here.

Last, we want this farm to be a place in which "civility and the intellectual and moral life can be sustained through the new dark ages which are already upon us." MacIntyre's language may seem a little apocalyptic, but surely the best way to wait for another—doubtless very different—St. Benedict is to pray and work, and "Benedict Farms" offers us plenty of opportunities for both.

I have dedicated this book to Andrea, without whom neither our farm nor these reflections would exist. There seems something a little inappropriate about dedicating a book to someone who is more coauthor than dedicatee. Nonetheless, it's a small reflection of my immense joy and gratitude for the privilege of spending my life with this woman I love so dearly.

A BURNT OFFERING

We live on a small farm in Central Texas. We raise some heritage livestock breeds, and our youngest son, Andrew, shows goats and cattle through his school. On Monday our goat died. We bought "Daisy" last September as a companion for "24," Andrew's show goat last year. Well, 24 made the sale at the County Show in February and thus went the way of all flesh, but we kept Daisy. She was playful and affectionate, and since 24 has been gone, Daisy has really become like one of the sheep. But she was also becoming a pain. First, she's gotten much larger, and she's much, much stronger. And second, her desire for food—everybody's food (sheep, chickens, geese, guineas, etc.)—was just overpowering. She was also starting to get pretty aggressive. (And her horns grew faster than we could trim them.) Last week she broke through another fence and tore down a tree that we had planted on the anniversary of Andrea's mother's death. We decided she had to go.

And then on Monday, she broke into one of the chicken coops and got her horns tangled up in the chains holding the chicken feed that we raise up high during the day (to keep it away from the lambs who are still small enough to come through the open chicken door). When we got home, I found Daisy strangled and hanging from the feed bucket and chains. I didn't realize at first that she was dead. I was trying to free her, but she was so heavy that I couldn't get her horns free. And then I realized that she wasn't fighting me. And when I did get her free, she simply slumped to the ground.

Well, I wasn't terribly upset about this, but I certainly didn't want it to end this way. And Andrew was quite attached to her. She was always most affectionate toward him. She followed him around like a big dog. I gave the "farm animals aren't pets" speech again, and amidst his tears, the three of us loaded her into a wheelbarrow and then into a big trash bag.

So now what do I do with a 150-pound dead goat in a trash bag? Burying her would take hours. I thought momentarily of trying to hide her in a big trash can and hoping that the trash people (who come on Thursdays) would take her away unawares. But this plan never had a chance. By Wednesday morning every fly in the county was swarming around the trash bag, and the smell was atrocious.

And thus tonight with some difficulty I managed to move her, the flies, and the smell to a burn pile in the pasture. Her trash bag was now filled with goat bodily fluids, and it was quite an effort to lift her on to the top of the pile. The bag was ripping, the fluids were going everywhere, the smell was beyond words, and the flies were having a heyday. I smelled as bad as the goat. I finally got her halfway up the pile. I poured kerosene on it and lit the fire.

What I didn't count on was just how long it takes a goat to burn. All my life I've read about burnt offerings in Scripture, but it's never occurred to me just how long it takes the animal to be consumed. Most of those burnt offerings in Scripture could have taken hours. And during that time, one does a lot of thinking. The winds had picked up, and I was battling the smoke and fire and the dry grasses in the pasture while trying to move the wood over the goat who simply would not burn very quickly. Her head and horns were among the last bits of her to be consumed. It took about three hours.

I had wanted to get rid of the goat, and now I got what I wanted. But I did not envision this. I burned a goat that was a cheerful annoyance in both life and death. She had no great value to the farm. My burnt offering was no real sacrifice.

How glibly I use the word "sacrifice." I use it to describe inconveniences and those choices I make among the many preferences I have for how I spend my time, money, and energy. In the ancient world, one sacrificed animals "without blemish." You did not get rid of your problems and call it sacrifice. You gave your best. What would it mean to sacrifice an ox? Three-quarters of a ton of meat and bone that would have been desperately needed to carry goods and people from one place to the next or to till a field in a parched, dry land? And you simply gave it to God?

How much I have to learn.

BUYING GEESE

A few weeks ago, I bought two Saddleback Pomeranian Geese that we named George and Gracie. And then a couple of weeks ago, a raccoon got in the pen and brutally killed Gracie. So, I've been looking for another Saddleback. I found a few online, but they were all really expensive and a long way away.

Well, this week I found some in Bryan and this guy was only asking twenty-five dollars each for them. This morning I went to Bryan to check them out, and I bought two geese and a gander.

Of course, now I had to get them home. I put them in a big dog kennel in the back of our Suburban, and as soon as I got them in the car they started crapping. I had put down a tarp under the cage and old feed bags around the sides, but the stuff was going everywhere. I was not five minutes away from this guy's farm when the smell in the car just became overpowering. I raised the back window and that helped some with the smell, but then the wind noise and the noise from the traffic was really loud. And the geese did not like it AT ALL. So they start honking to beat the band. And they're LOUD. At the time, the Rangers were close to making the playoffs, and all along I had planned to listen to the game on my return home. But to listen to the game on the radio and to hear it over the honking and the wind noise I had to turn it up very high.

So for 110 miles, I had wind noise, traffic, the radio blaring at full blast, and crazy loud geese honking and shitting like there's no tomorrow. And then of course it started to rain. I closed some of the windows, but the smell was just nauseating. They would have to stay open, and I would just "git wet, I reckoned." Frony's reply to Dilsey rang in my ears: "I ain't never stopped no rain yit."

When I finally got to our farm, I realized that now I had to get them out of the kennel and into the pen. The back of the car was a war zone. When I went to reach into the kennel to get the first one, they all went crazy—honking and scurrying and flapping their huge wings and sending goose shit absolutely everywhere. By the time I had gotten all of them out of the car and into their new pen, I was covered in goose shit and there were thousands of flies in my car.

The three new geese met George, and the goose dance began. With fresh water and some feed, they settled in quickly. We've named the new gander Fred and the two new geese Ginger and Gracie. Geese mate for life, so I am assuming that George and Fred will determine which goose is Gracie and which one is Ginger.

But how do you clean the inside of a car that is absolutely covered in goose shit? Several hours and rolls of paper towels later, the visible remnants were gone, and the car smelled a bit like an old hospital in which the overpowering odor of bleach almost but doesn't quite remove whatever smell it was trying to cover up.

UPDATE: It's been two weeks. I'm pleased to say that the geese seem to be doing quite well, and the raccoons have not yet returned. However, there are still flies in my car.

COMING TO TERMS WITH SHIT

One of the first things that we had to learn on the farm was how to think about shit. Actually that's not quite right. The first thing we had to learn was how to talk about shit. There are of course lots of possibilities when it comes to naming excrement: manure, crap, poop, feces, dung, turd, number two, and on and on. And when we start to make distinctions based on the animals who are doing the defecating, we get more terms. Horses produce road apples, poultry guano, and cattle cow patties, pies, meadow muffins, and (when properly dried) buffalo chips. But most farmers just refer to all of this as "shit." And it's not a cuss word. It's just shit. Few things in life are as important as learning to call things by their proper names. If Harry can call Voldemort by his real name, we ought to be able to do the same with shit.

But "shit" is a bad word. Polite people don't say the word in public, and they certainly don't discuss the subject matter, or at least not while sober. Some people use it as an expletive, which I don't really understand. Perhaps when in anger or frustration someone says, "Shit!" then that is just shorthand for "this _____ is like shit!" But I have also overheard my students say, "This is the shit!" and they meant it as a compliment. I don't have any idea what that means. Wittgenstein reminded us that words don't have meaning so much as they have usage, and that is clearly on display here. He also said that sometimes language just goes "on holiday," and that probably applies here as well.

On any account most city people are unnaturally terrified of shit. They don't want to see it or smell it, and heavens to Betsy, they don't want it to touch them or anything they own. They typically only encounter it in two forms: their own and that left by neighborhood pets, especially dogs, cats, and birds. Concerning the former we go to great (and understandable) lengths to get rid of it as quickly as possible, certainly without touching it and ideally without even seeing it. We buy expensive sprays and perfumes to remove all traces of its presence. Concerning the latter many cat owners use litter boxes, and dog owners take Fido for a "walk," but properly dispensing with excreta is now a high moral obligation. Most cities (and virtually all apartment complexes) have vigorous laws prescribing what you are supposed to do with your dog's droppings, and in lots of places you can even get a ticket or a fine for failing to do your duty. And most city people who step on a dog pile are mortified (see the expletive explanation above).

But why all the fuss? Is there anything in the world—I mean literally anything in the world—more common than shit? Given the craziness of what and how people eat these days, breathing and shitting may simply be the two things that we all have in common.

I want to be careful not to glorify or idealize it because sometimes it's pretty revolting. But the point is that on a farm you simply can't avoid shit. It's everywhere. Sheep produce hundreds of little marble-sized bits of dung (think Junior Mints or on occasion Milk Duds). In the barnyard corral where the sheep stay at night, sheep shit is everywhere. It's impossible to walk through that corral without stepping on hundreds of little Junior Mints. It dries quickly and doesn't really smell, so it's no big deal. But we've had lots of guests come out to the farm, and many will tiptoe through the corral as if they were walking through a minefield.

Cow patties are a little more problematic, and you do want to avoid stepping in fresh ones if you can, but you learn early on that they will be almost impossible to avoid entirely. And the same goes with the chickens and the geese and the turkeys (oh man, the turkeys). If you're going to work with animals, you're going to get shit on your shoes—and your pants and your hands, and if you're not wearing a hat, probably in your hair as well. You simply have to come to terms with shit.

Sometimes our encounters with shit are nothing short of spectacular. Earlier I told the story of transporting some new geese in the back of our Suburban in which I was covered in goose shit when I tried to retrieve them from the back of the car. There are numerous such stories. I do believe that every time a cow gets into a trailer, the first thing she does is drop a load. It took me years to figure this out. There I would be, loading a couple of thousand-pound-plus heifers into a trailer, trying to get them lined up and tied into their respective spots. It's always a pretty tight fit in our little trailer, and while I'm negotiating the ropes and the cattle and the doors and trying not to get stepped on, one would back me into the side of the trailer and either baptize me or drop a load on my jeans and boots. The first few times this happens, it's pretty unsettling. But after awhile, you can only laugh at yourself because you didn't see it coming (again).

And here's the great news: it's really not so bad, and everything cleans up. Sooner or later you quit worrying about the shit, and you simply go about your farm chores. Yes, from time to time, Andrea and I have become so comfortable with it that we have failed to notice that we were carrying evidence of earlier excreta back into town with us after the chores were done, but on the whole, we clean up pretty well.

That's another difference between city and country people. City people bathe to get ready for work (usually in the morning), and country people bathe after the work is done (before bed or before the evening activities). It makes perfect sense. If I know that I'm going to be shoveling shit cleaning out a barn, it doesn't make a lot of sense to get cleaned up beforehand.

The most important lesson in all of this is of course that shit is also enormously useful for the garden, fields, and compost piles. It has to be managed and used in proper proportion with other types of organic matter (there's another lesson here), but even the most novice gardener will recognize how beneficial it can be to the long-term health of the garden and the fields. The best book I know on the subject is Gene Logsdon's *Holy Shit: Managing Manure to Save Mankind*. Logsdon is at his witty and insightful best when describing the wonderful world of shit, breaking it down into categories, and explaining what types are best used in what ways. What's the difference between "hot" and "cold" manure? Herbivore versus carnivore shit? How best to preserve or diminish high nitrogen content? Can cow methane really blow up your barn? Logsdon knew his shit. 'Tis a consummation devoutly to be wished.

LAMBING

All afternoon we could tell that this ewe was in labor. She stayed in the barn and wouldn't go out with the others. She had quit eating and seemed to spend most of her time trying to get comfortable. I had been afraid that maybe her lamb had died inside her. She should have given birth long before now. The other ewe (her mother actually) had given birth to her twins long ago, and they were both exposed to our ram for the same amount of time. And here we were two months later and still no lamb.

But after dinner Andrea and I went back out to check on her, and it was really quite dramatic. Her breathing was labored, and she still couldn't get comfortable. We checked on her off and on all evening. About ten o'clock it was clear that something new was happening. We had read (and watched YouTube videos) that we should see a nose and two front hooves make their appearance first. The experts were clear that if you did NOT see two hooves, then it was imperative to find the other leg and turn it around so that it did not break during delivery.

At first all we saw was a nose and one hoof. Attempting our best James Herriott impersonation, Andrea went first. Her hands were smaller, but she couldn't find the leg. I eased my hand along the body of the yet-to-be-born lamb, feeling my way to the little leg. I found it and turned it around. True to form, the body of the little lamb slowly began to come forth. While Andrea ran to the house to get a towel and iodine, I watched in amazement when a second lamb seemed to slide right out five minutes later.

After the births we spent the next few hours watching and occasionally trying to help them nurse. The firstborn was a little ram, and he got up on his feet pretty quickly. The second was a little girl, and she was having a lot of difficulty. She was very small, and it took her about two hours to stand, and then she could only nurse when Andrea would hold her up and put the teat into her mouth. By the time we left, she could stand and nurse a little but not much. We hoped she would make it through the night. Nature is extraordinary, but it's also brutal.

I don't know that I have ever experienced spring quite like this. In addition to the lambs, we have eleven baby goslings and fifteen ducklings. The mother duck has taken the ducklings somewhere. We haven't seen them for several days. We did find one gosling that apparently hatched late and wandered into the garden, shivering in the corner of a raised bed. When we tried to reunite her with the other geese, they rejected her and chased her away. So we're raising her separately. She was born on Holy Saturday, and Andrea named her Easter. We bought her a companion duckling, and they are in a water trough in the coop with the guineas. We'll see.

GUINEA FOWL

Guinea fowl are almost the perfect farm animals. Adult birds require almost no attention and maintenance. You don't even have to feed them, though they love a little chicken scratch or millet. They eat all sorts of slugs and bugs and ticks. When the rest of the county is complaining about the annual hordes of grasshoppers and crickets, those of us with guineas barely notice. And while they eat bugs, they do not usually eat fruits and vegetables. This means that you can let guinea fowl, unlike chickens, into the garden without fear that they will devastate your tomato plants or squash or peas or beans. And their guano is compost gold.

They are superb watchdogs. If anything (or anybody) shows up on the farm that's new or not supposed to be there, they go bananas. Squawking and fussing, if there's something new on the farm, the guineas will tell you about it. They hate snakes and, working together as a flock, will corner and kill most species. They are fearless. I once watched our flock surround an opossum and drive him off of our farm and across the road into Hog Creek. They returned squawking triumphantly. They have even been known to drive away rattlesnakes.

And, if one were so inclined, they can of course be harvested. Guinea fowl is a delicacy often served in the finest restaurants. And their eggs are quite tasty, though the hens typically only lay from March to September.

Unfortunately guinea fowl are only *almost* the perfect farm animal. They do have some deficiencies, and the most significant of these is that guineas are unfathomably stupid. They quite literally do not have the sense to come in out of the rain.

Imagine that one or two guineas get separated from the rest of the flock by finding themselves on the wrong side of a fence. There will be no peace on the farm until they are reunited. They will run up and down along the fence, squawking at full volume. Typically the flock will respond to the absence of these two by joining them in their long and loud lament, longing for their companions to join them on the right side of the fence.

As if on cue, the entire heavenly host of guineas will run back and forth on both sides of the fence, vigorously decrying the injustice of the situation. These birds can fly, but it will rarely occur to them simply to fly over the four-foot fence that separates them all. On numerous occasions we have attempted to herd them toward an opening in the fence only to see them repeatedly *run past* the opening all the while still squawking at the top of their lungs. We've seen them spend two days and two nights on the wrong side of a fence, and no amount of trickery or persuasion is guaranteed to get them through the gate.

Moreover, guineas are wanderers (which is why they get separated from one another). In the best-case scenario, they will stay fairly near the house and garden, meandering here and there, eating grasshoppers and chasing off snakes. In the worst-case scenario, they will wander into the road, or down the road, oblivious to the dangers of traffic. Most birds scatter in the face of an oncoming car. The guinea *might* scatter as well, but he's just as likely to look up at the car, squawk and screech, and get run over. We've seen our flock—both the quick and the dead—half a mile from our house, oblivious to where they were or what they were doing. This proclivity to wander is the second-biggest reason why you can't reasonably have guineas in the city or even in the suburbs. They will drive your neighbors crazy.

Of course, the biggest reason why you can't have guineas in the city is the issue of sheer volume. They are loud, plain and simple. You think a couple of roosters might annoy your neighbors with their all-day wake-up calls? Try two dozen guinea fowl who NEVER STOP SQUAWKING. They put the roosters to shame. Above I said that if guineas discover something new on the farm, they will sound the alarm. That's true, but sometimes the something new is a shovel that's been left out or a towel blowing in the breeze. They are not especially good at knowing what's new and what's not. And so they tell you about all of it—all the time.

And for all of their fearlessness, their stupidity so often trumps their bravery. They simply don't recognize some predators. The hens make their nests on the ground in high grass, and when they are sitting on a nest, they are extraordinarily vulnerable to predators, both large and small. And at night guineas like to roost in the trees. Owls are one of their most dangerous predators. The owls will come and sit beside the unsuspecting guineas on a limb and then bite their heads off. On many a morning, we've found a headless guinea lying lifeless beneath the tall hackberry tree where they like to roost.

The solution to this problem (so we have been told) is to train the guineas to come into a coop in the evening where they will be safe from predators. Supposedly if they learn that they will get a little seed at sundown, they will return every night for their treat. The idea is that when they are still keets (guinea chicks), you feed them in their coop. After you let them out of the coop, you only scatter seed for them in their coop. Ideally they will come back on their own. But we have had no luck with this plan. So we're back to the task of trying to herd them into their coop. (See guinea herding above.)

Consequently very few of our guineas last for more than one year. Last year we bought two dozen, and only four made it to the one-year anniversary. But they're worth the risk. In the one year we didn't replenish our guineas, we were overrun with grasshoppers. They annihilated the leaves on the trees. They were so thick in the high grass that when we mowed, they would fly up and into our faces like swarms of bees. That summer we learned our lesson. Better dumb guineas than no guineas at all.

RED IN TOOTH AND CLAW

Our farm education continues.

Last week a couple of dogs got into the chicken pen and coop. It was devastating. When we got there in the evening, we initially thought that they might have gotten all of the hens—feathers and dead chickens were scattered across the yard and the coop. After chasing the dogs off, we discovered that there were some chickens hiding; others were badly wounded but still alive. Final count: five dead, two so badly wounded that I had to kill them, four significantly injured, and four mostly unharmed. I don't know if the four walking wounded will make it.

Andrea was bound and determined to nurse the wounded back to life. She created a poultry infirmary for them out of some dog kennels. With the chickens separated and in a secure spot in one of the stables, Andrea poured hydrogen peroxide on the worst of the wounds and gave them some feed and plenty of fresh water. It's almost been a week, and they're hanging on.

Of the two I had to kill, I got some firsthand experience with the old phrase "running around like a chicken with its head cut off." In my case after I thought I had killed the first one (and put her in the trash can), she was flopping around amidst all the feathers and chicken parts. I had to reach in, bring her out, and "kill" her again. Back in the can and at that point completely headless, she was still flapping around. Number two put up even more of a fight—which seemed outrageous since all of her organs were hanging out, and she could hardly walk. But the deed was done.

The next day the dogs were back. And from their perspective, why not? What dog wouldn't want a fresh chicken dinner? We chased them away again. Friends have told us that we can shoot them if they are threatening our livestock, but we're not sure about the law, and even then I don't know if I could do that.

I drove down the road and found both dogs on the porch of an old farmhouse. I called out from the yard, and the owners came out. We had a good chat. At first they said their dogs didn't eat chickens, and I said, "Well, they did on Sunday. I chased those two dogs right there out of the pen myself."

I told them that I understand about dogs and that we have dogs ourselves. "But it's important for you to keep your dogs penned up. I wouldn't want anything to happen to them, but if they come back, we will have to protect our flock." They apologized and said they understood.

And then I told them that they had some mighty fine-looking horses.

PLAYING BY EAR

Anyone who does much language study knows that training your ear to hear the language spoken correctly is half the battle of language acquisition. If you know what sounds right, then when you encounter the words spoken incorrectly, they will *sound wrong*, and this is far more effective than merely trying to memorize a bunch of rules. We spend a lot of time learning grammar rules, but this pales in comparison to being in an environment in which one hears the language spoken properly. And of course the converse is true as well. If we are in an environment in which we hear and speak words and phrases incorrectly, then we will have little or no basis for hearing the error. In this context what's right *sounds wrong*, and it's very difficult to correct it.

When I was in graduate school, one of the many part-time jobs I took was to teach "business English" at a nighttime secretarial school. Almost all of my students were young to middle-aged women who had dropped out of high school and were now trying to get the skills necessary for them to work in an office or clerical job. One day we were talking about answering the phone, and I was trying to help them make the transition from saying "This is her" to "This is she." All talk of nominative and objective was lost here. I was trying to explain how important it is to say the right thing over and over again, so that it will sound right. And then I made one of my many mistakes. I said, "In the same way, if you knock on the door and someone says, 'Who is it?' the correct response, even though it may sound wrong, is 'it is I.'" There was a young woman sitting on the front row who chimed in, "If I said 'it is I,' they wudn't know it was me."

We train our ears to hear what the farm sounds like as well. And when there's a problem or something unusual has happened, we often hear it before we see it. Most of the time, all we know is that "something is different," and we go in search of the explanation. The tin on the roof of the far side of the barn is starting to come loose, and we hear an unusual banging. Weak tree limbs start creaking before they fall, and the sound of running water that is not supposed to be running is unmistakable.

All the animals have their own sounds, and they can bring both good news and bad. If a lamb has been born overnight, the chance is good that you will hear a new sound, a high-pitched bleating that wasn't in the corral yesterday, before you discover the new little fellow. On many occasions we have heard the unmistakable sound of baby chicks that have hatched from some unknown nest before we find them. These are wonderful discoveries.

But the sounds also bring bad news. The most important of these are the moments of danger. An animal caught in a fence or tangled up in a rope or wire brings a scary call of distress. The heifer we thought was bred tells us that she is not long before the vet confirms this disappointment. When all hell breaks loose with a predator in the chicken coop, you hear it before you discover the carnage. And of course there is the dreadful blessing of the diamondback's rattle.

Dogs have different barks. There's the excited bark of seeing a rabbit, the long-distance bark of barking at (to?) other dogs on neighbor farms, the boring bark of warning the geese, and the menacing bark of confronting a snake, an opossum, or a coyote. Oftentimes late at night when I am inside reading, I hear a "different bark" that leads me to go outside and see what's going on.

Similarly, cows have different kinds of mooing. As mentioned above, a heifer in estrus will usually make quite a demonstrative presentation. Long and loud, mournful and melancholy, she will walk up and down the fence line lamenting her open (unbred) state. This mooing is quite different from that of the mama cow who has been separated from her calf. As part of our Beefmaster breeding program, we have to weigh the calves within the first day of their birth. Taking that calf away from Mama—even for only a few minutes—will produce a very different sort of complaint. And of course there are a variety of other kinds of mooing, most of which we don't understand. But the more time you spend with cattle, the easier it is to detect bored lowing from plaintive demands.

Engines and equipment also have their own particular sounds. When you use them a lot, you know how they sound—or at least how they are supposed to sound. Tractors, trimmers, saws, drills, sanders, air compressors, power washers, and all the rest each have their own sound. It is on the basis of how it sounds that we know we must attend to it, and it is not infrequently the sound that tells us that we have (or have not yet) successfully corrected the problem.

But of course none of this can happen unless you are constantly attending to the farm and learning how it sounds ordinarily. Once again the habit of attention is the key. I love music and have thousands of songs saved to various iPods and phones, but I don't ever wear headphones when I'm doing farm chores. I might listen to a game or the radio while working in the shop but not in the pastures, the corral, or the barns. Mostly it's because I need to focus on the task at hand, and I don't want to be distracted by the music. But the real reason is that I love the sounds of the farm, and real love is inseparable from learning. I like to hear the sheep bleating, the roosters crowing, and the hens clucking. I love (as E. B. White describes them) the "garrulous geese." I remain enthralled by the snorting, grunting, and squealing of the pigs. And this is to say nothing of the many songbirds that come through Central Texas: sparrows, swallows, thrushes, orioles, and so many more, each of which make their own contribution to the barnyard chorus. And when those sounds are replaced by new and unexpected noises, well . . . we learn to say with Clara and Miss Clavel, "Something is not right."

One of the great surprises and joys of the farm is how much you learn by listening. Perhaps this is a principle we should apply to other areas of our lives as well. How much more of our world will we appreciate when we learn to play by ear?

HOMECOMING AND THE FUTURE OF HIGHER EDUCATION

In Wes Jackson's remarkable 1994 book *Becoming Native to This Place*, he notes that most colleges and universities now offer variations of only one major, upward mobility. What was true twenty years ago is even more the case today. In part to justify the explosive costs of higher education, students are directed toward ever greater specialization and expertise to enable them to move out and to move up. Jackson continues, "Little attention is paid to educating the young to return home, or to go to some other place and dig in. There is no such thing as a 'homecoming' major."[1]

Many might dismiss Jackson's suggestion as nostalgic fantasy. Why should one pay $50,000 a year merely to go home and "dig in"? Of course one should not. But it is not unreasonable to suspect that the day is coming when students and their families may not be willing or able to continue to pay (or borrow for) the ever-increasing and outrageous fees that we charge, especially when so often they get a view of the world that is deeply misleading and sometimes manifestly false. As academic departments many of us continue to indulge ourselves in the belief that our "disciplines" are autonomous kingdoms of segregated knowledge, the mastery of which will justify the fees and guarantee the upward mobility our students and their parents so ardently desire. We specialize in the credentialing and certification of this mastery.

Of course, most of us are not lying. We actually believe in the disciplinary fragmentation of knowledge and in the superiority of the special pleading we call our methodologies and procedures. In recent decades most of us have also become devotees of a high-powered technological messianism in which a new app, method, or machine will solve all or most of our problems. And our students are willing disciples. Their expectations are exceeded only by their confidence that their college degrees will insure them against poverty and want. Unfortunately it is not true. Many of our students lack the wisdom, the knowledge, and the skills to flourish as they might, and we will not get better by merely fine-tuning our instruments of assessment. How then should we begin to think about these matters?

What if colleges and universities were to recognize and begin to enable our students to flourish in a world of coming scarcity? Is it possible that we might direct some of our energies toward equipping students with not only the skills but also the rationale for curbing expectations and adapting one's life to meet what is available in a given place at a given time? What Jackson calls a "homecoming" major is really a retrieval of a series of transformative practices—most notably sustainable agriculture, cottage farming, community development, home economics, moral philosophy, and literature—for the flourishing of the life of the mind and the cultures of civilization.

One of the great ironies of contemporary higher education is that STEM and the humanities have their intersection at a place that both claim to have "outgrown," namely their agrarian heritage. Colleges and universities must retrieve and reclaim this agrarian past as the key to an uncertain future. By reclaiming and refocusing our attention toward the integrated knowledge, habituated practices, and virtuous dispositions of our agrarian origins, we might find new resources for navigating the troubled disciplinary waters of higher education.

How might we reclaim and retrieve this agrarian past? At the most radical level, it would require rethinking the entire enterprise of higher education. Most of us and our institutions simply cannot do this. However, we can begin to take steps that will place our institutions and our students on "more fertile ground." I am not really proposing a new "major," nor am I suggesting that merely adding one more healthy option to the already bloated curriculum buffet can solve all our woes. But I believe that we must take steps to reorganize and reshape our cultural and curricular focus in a way that privileges the development of sustainable agriculture within settings of cottage and urban farming. We are simply going to have to learn not only to feed ourselves but to work with our neighbors on recreating communities of mutual dependence and to understand those goods that make human flourishing possible.

How would we do this? There are numerous ways, but as it stands right now, too few of us have turned our attention in this direction. It might be through creating interdisciplinary programs or establishing new schools and initiatives that will equip existing departments to pool their resources. We must figure out ways to bring university costs down while equipping students with both integrated knowledge and skill sets. I know of no place where this can be more beautifully and effectively achieved than on a farm, be it a small plot of acreage in a rural setting or a collection of greenhouses and raised beds on a Manhattan rooftop. And as Aristotle taught us long ago, the integration of knowledge and practice will lead to wisdom.

THE BENEFITS OF AN AGRARIAN EDUCATION

At the curricular level, one of the most important benefits of retrieving an agrarian orientation is the explicit dissolution of the fragmentation of knowledge into disciplines. Life on the farm teaches the unity of knowledge. Economics can only be divorced from biology at our gravest peril. Every barn, building, and breed is lived history, and these histories are embodied in the stories of longing, failure, and occasional success that constitute our lives. These failures and successes are largely the product of how well we have learned our lessons and how closely we have paid attention. There is art and beauty at every turn. Composting is merely the purposeful redirection of the material excesses of sociology toward the production, preservation, and conservation of topsoil. There is no disciplinary fragmentation on the farm.

A remarkable catalogue of benefits follows here. The most obvious, of course, are the benefits to the sciences. Students and faculty who are working and studying in an agrarian context will naturally have substantial engagements with biology, geology, zoology, physics, mathematics, chemistry, environmental science, ecology, botany, animal husbandry, entomology, soil science, ornithology, landscape design, mechanics, astronomy, and meteorology, to name only a few. And this is to say nothing of the interaction between the natural and the social sciences. The politics and the economics of the cottage farm bring one into sharp disagreement with the conventional "wisdom" of modern industrial agriculture. Students will not study any of these fields or disciplines in isolation, let alone in competition with one another. Such a context shows the folly of much of the abstraction of modern higher education. The agrarian context also offers excellent resources for an informed evaluation of the proper uses of technology.

But it is not just a benefit to the sciences. What distinguishes the farm from the science laboratory simpliciter is that here there is no charade of objectivity. The farm is the place where all that science is put to use by and for human beings, amidst their loves and losses, their trials and their triumphs. This is the place where our stories begin, and it is entirely fitting that some of the greatest literature in human history has its setting and its focus on the life lived amidst the cultivation of the fields. This is one of the essential themes of Western intellectual history and literature. From Virgil's *Georgics* to Willa Cather's beloved Nebraska, from the lyrical ballads of Wordsworth's Lake District to the unforgiving realism of Thomas Hardy's Wessex, from the cultural politics of the Southern Agrarians to the hilarity of Stella Gibson's *Cold Comfort Farm*, in this literature we see the beautiful and the tragic embodied in the experience of the natural world. Every language and culture has its representatives of this universal experience, and it is the poems, the songs, the art, and the music that celebrate the diversity of the natural world in the diversity of the many languages and cultures in which it is embodied.

But there is more to it than that. Those who know me will find it no surprise that I believe that there are no two vocations that are so mutually edifying and reinforcing as those of the philosopher and the farmer. Aristotle said that philosophy begins in wonder, and the farm is nothing if not a school for wonder. There are epistemological problems: "What will happen if I . . . ?" There are conceptual and linguistic problems: "How do I describe what I am observing?" There are moral problems: "What chemicals, if any, can I use on these crops?" "Do animals feel pain?" "What are my obligations to these creatures and to this land?" There are practical and prudential problems: "What do I do with a 150-pound dead goat?"

It should go without saying that there are also many benefits to society, not the least of which are fresh vegetables. Students schooled in this way will bring habits, insights, and skills for sustainable living in a coming period of scarcity—and there will of course be a coming period of scarcity. We live in an era of unrestrained consumption, and students who have lived and worked in this way will appreciate the value of frugality and will understand the essential values of conservation and enhanced ecological consciousness. They will understand that convenience is not a right and that those goods that will sustain human communities will require modesty, mutual cooperation, and perseverance.

Of course, I think that the greatest benefit comes to the students themselves. The interplay of productive work and meaningful leisure found in this agrarian context provides them with an engaging model of *eudaimonia*, or human flourishing and happiness. Such *eudaimonia* is of course the natural *telos* of human beings, but it is only possible through the *cultivation* of the moral and intellectual virtues. If they are ever truly to be happy, if they are ever to find and build and sustain a home of their own, they will have to cultivate these virtues. It is of course no accident, that virtues—like vegetables—require cultivation.

CAN COLLEGES ACCOMPLISH THIS?

Yes, I believe that they can. David W. Orr has argued that the modern agricultural dilemma of separating the study of agriculture from its communal and ecological contexts began when "agricultural sciences were isolated in research institutions and from there evolved into technical disciplines whose purpose was to do one thing: increase production. Consequently they were not rooted in any coherent and sustainable social, philosophical, political or ecological context ... [and furthermore] a great many assumptions about nature, technology, farming, rural life, and the consequences of applying industrial techniques to complex biological and cultural systems went unchallenged."[2] According to Orr, it might have been very different if agriculture had evolved within liberal arts colleges instead. "In the context of liberal arts colleges, agriculturalists might have learned to see farming not as a production problem to be fixed, but as a more complex activity, at once cultural, ethical, ecological, and political."[3] Orr quotes Aldo Leopold, author of the magnificent *Sand County Almanac*, as believing that the goal of liberal education is "not merely a dilute dosage of technical education" but rather "to teach the student to see the land, to understand what he sees, and enjoy what he understands."[4]

Most colleges and universities are better equipped to address this need than they realize. Despite the fact that research and teaching in agriculture is now almost the exclusive province of land-grant universities that are often deeply beholden to industrial agriculture and the agribusiness conglomerates that fund the research and pay for the facilities, there is no reason why smaller schools could not turn some of their natural and social science resources toward the development of cottage and urban farming initiatives. Many small state universities and some liberal arts colleges work vigorously to serve the rural communities that surround them. But many schools are tempted to buy into the belief that future agriculture success will be based on the notion that the farmer and the college must either "get big or get out." Such need not be the case. While small liberal arts or religious colleges do not usually have the resources for cutting-edge STEM research, they do have fine programs in the sciences that are complemented by a serious and vigorous commitment to the humanities and liberal arts. And these are precisely the sorts of places where students can learn to see the land, to understand what they see, and enjoy what they understand.

This is especially the case for faith-based institutions, places where the integration of faith and learning is not only not excluded but also given the place of prominence and direction. Religious colleges and universities have the benefit of being able to appeal to a doctrine of creation that engenders a robust account of stewardship that transcends any mere utilitarian account of conservation. As Joel Salatin has recently argued in his wonderful *The Marvelous Pigness of Pigs*, it is a great tragedy that so many Christians and so many environmentalists think that they ought to be opposed to one another when in reality they share so very many foundational commitments.[5]

At one time (and still at many smaller regional state universities), colleges maintained a "college farm." Schools as diverse as Dickinson College in Carlisle, Pennsylvania, and Southern Arkansas University in Magnolia, Arkansas, show some of the possibilities that can be achieved. David Orr lists seven benefits to the establishment of a college farm.[6] (1) It offers a unique experience that is no longer available to most students, the majority of whom have not grown up in rural environments. (2) It is to be an interdisciplinary laboratory for the study of the numerous fields discussed above. (3) College farms can "become catalysts in the larger effort to revitalize rural areas." (4) They can preserve that biological diversity that is jeopardized by both industrial agriculture and urban sprawl. (5) They can reduce carbon emissions involved in the long-distance transport of food. (6) They can "close waste loops by composting all campus organic wastes." (7) By participating in the design and maintenance of the college farm, students learn to take responsibility for local problem-solving and decision-making.

I would add to these reasons that the college farm can become not only a teaching laboratory for credit classes, but also the source of work-study aid to students. It is a fellowship of sorts. There can be individual research projects as well as the regular cultivation of a significant source for the institution's own food service needs. This would also allow for sustained interaction with area farmers and rural residents, further enhancing the relationships between town and gown, or rather, in this case, between the undergrads and overalls. "Farm-to-table" is all the rage these days. Why not "farm-to-dining hall" or a student farmer's market?

It just might be that herein lies the basis for a thriving program in sustainable agriculture, environmental science, and agrarian literature, philosophy, and history that attempts to unify and integrate fragmented knowledge and reduce tuition and costs in the process. Armed with such insights and skills, the future of higher education may not be so bleak, and some of our students might just make it home after all.[7]

DOING IT AND GETTING IT DONE

When it comes to farm chores, few distinctions are as valuable as learning the difference between doing a job and getting it done. Nothing will take the joy out of farming more quickly than merely trying to finish your chores. When I'm running late, or if it's too cold (or too hot, or too wet), or anytime I just want to "get it done" so that I can go do whatever it is that seems more important or more interesting to me at the moment, then there is very little joy in farming. Joy, not to mention success, comes from doing a task rather than just trying to get it done.

And of course this applies to virtually any task (not just farming). We have to learn to enjoy the work that we're doing, to give it the attention and the time it requires, and to take pride in the work itself. It's so easy to be bothered by the many tasks we have before us, but if we simply slow down, see what needs to be done, and set ourselves to the task of doing it, it's almost always more enjoyable. (Andrea and I sometimes disagree on the "slow down" part; she believes, surely with some justice, that I could often pick up the pace and still not miss out on the beauty of the moment. She says slow and careful doesn't have to be "slow motion.")

Wendell Berry remarks on more than one occasion that there are only two reasons to farm: because you have to or because you love to. I know many farmers that fall into the "have to" category. They are members of families that started down this road long ago, and truth be told, they have few other options. This is what they do. Pursuing other options would mean making matters difficult for other family members; there is just too much invested in the land and the equipment and the family. They farm because they have to.

The ones who are happy also farm because they love to. How most of them make ends meet is beyond me.

Silly hobby farmers like us start out doing it for many reasons, but most of these ultimately boil down to love, either real or imagined. We love the open spaces and the fresh air; we fantasize about self-sufficiency, independence, and "the simple life." We're tired of urban nonsense and want to try our hand at some rural nonsense for awhile. Most of us want our children and grandchildren to experience a world that is quickly slipping farther and farther away from the urban and suburban lives they have come to take for granted.

But sentimentality and idealism rarely have the power to sustain us through difficult times, and I can't imagine that there's any reason for people like us to keep on farming except for love. And of course some days it's hard to love. It's difficult, messy, and expensive. And on a cold, rainy morning when the mud and muck are so thick that one boot gets stuck in the mud, and your sock foot comes all the way out of the boot and lands in a puddle of mud and manure just before a cow knocks you into another muddy puddle, you easily find yourself wondering, "What am I doing this for?"

At that moment, it's best not to make any long-term plans. The best thing is to smile, look around, and see what needs to be done. Be glad you wore your old clothes, unlike the other day when you stopped by the barn to check on one little thing before heading to work in a coat and tie only to have to go back inside and change clothes after you carelessly ruined suit number one. When you're just trying to get it done, you're in a hurry and you wear the wrong clothes. If you're worrying about keeping your clothes and shoes clean, you don't notice the goose who is about to take a plug out of your leg or the ram who will knock you into the next county or just about anything having to do with the pigs. Wearing the wrong clothes makes you pay attention to the wrong things.

Of course, the other problem with merely trying to get it done is that you're doomed to failure. On the farm it's simply never all done. Even after everybody's been fed and watered, after the eggs have been collected, the seeds planted or the vegetables gathered, there is always more to be done. There's a fence or a barn roof that needs mending, a coop or a stall to be cleaned out, beds to be weeded, a wood shop to be cleaned up, tires to be patched, hooves to be trimmed, oil to be changed, and on and on. If happiness on the farm depends on getting it all done, then it's hopeless. But if we can find joy in doing these many tasks, then we do the things that must be done today and get to the rest when we can. This is a great consolation and why one can never be bored on a farm.

There aren't many rules here: if you don't know what you're doing, talk to someone smarter than you; start early; go slowly; pay attention; enjoy what you can; and be grateful.

The task is to do it, not to get it done.

THE FALLACY OF ACQUISITION

Most of us fall prey to the fallacy of acquisition. This is the mistaken belief that by acquiring something (a new car, computer, clothes, job, cell phone, etc.) I will solve the most pressing problem in my life. The fallacy of acquisition cultivates consoling fantasies about what is missing in our lives. "If only I had that _____, then all my problems would be solved." This is never true. The fallacy of acquisition is just another means of self-deception, and as I tell my students over and over again, the lies we tell to ourselves are so much more dangerous than the lies we tell to others.

On the farm the fallacy of acquisition grows exponentially because having the right tools and equipment really can make a huge difference in managing even the smallest of acreages. The most dangerous lies always contain some truth, and we latch on to that truth and ignore the rest. There are some tools that you really do need. But people lived on farms long before Tractor Supply arrived to tempt us with the latest gadget, and with a little planning and skill, most jobs can be completed without an extra trip to the store.

When we first purchased the farm, the only lawn mower that we had was a push mower that we used to take care of our suburban lawn. We bought our farm in the summer of 2011, which was one of the hottest and driest summers in Texas history. Everything was burned up, and we didn't really need a lawn mower until the next spring. But when spring arrived, all of a sudden the grasses around the house were knee-high.

Okay, that's not true. It wasn't all of a sudden. We watched it happen, but I didn't do anything about it until it was out of hand. We didn't live at the farm, we did not yet have any animals, and we were there mostly on the weekends. And when we were there, we spent most of our time renovating the small farmhouse. I kept telling myself, "I need to bring the lawn mower out here and clean up the yard." But I still needed the lawn mower at home (i.e., suburbia), and we didn't own a truck or a trailer, so it meant taking the handle of the lawn mower apart and putting it in the back of our SUV, only to put it all back together again before repeating the process to bring it home. In short, it was a hassle. And then it was too late.

When I finally got around to mowing the yard at the farm, it was a real struggle. The high green grasses stayed wet, clogged up the push mower, and in the worst spots, I would push it forward a couple of feet, have it give out, pull it back, clean it out, and do it all over again. It took all day, and when I was done, the yard hardly looked any better than when I started. It was clear to me that we had to get a riding mower.

We bought our first riding mower off of Craigslist, and we spent almost as much money trying to keep it running as we did buying it. Obviously we needed a *better* and *more powerful* mower. We upgraded to a new "lawn tractor" from John Deere, which has been truly wonderful for the couple of acres surrounding the house. But as for taking care of the pastures? I've been eyeing the larger tractors for several years.

And this is how the story goes. You can haul lots of lumber (and the occasional geese) in a Suburban, but sooner or later, you realize that you've got to have a truck. And that entirely adequate two-wheel drive truck is wonderful until it gets stuck in the mud. Then you start longing for a four-wheel drive one.

And tools? Oh my goodness. We are always one essential tool from having just what we need. A taller ladder, a special saw, all manner of pliers and wrenches and ratchets, not to mention the circular evolution from hand tools to electrical tools to battery powered and back to hand tools. Of the buying of tools, there is no end. And in all of this, the fallacy of acquisition silently guides us on. We look for things to buy rather than talking with neighbors about what else might be done or from whom we might borrow the tool we would only use occasionally if we bought it.

Of course, none of this is news. It was already an "old saying" when Lady Philosophy reminded Boethius long ago that "he who hath much, wants much."[1] Or as one of Wendell Berry's characters, Monroe Miller, put it, "To be satisfied with little is difficult. To be satisfied with much is impossible."[2]

MULES

The first animal that we brought to the farm was a seven-month-old mule. We bought her for $100 off of Craigslist.

We had a soft spot for mules already. In recent years we had become reacquainted with Andrea's mother's alma mater, Southern Arkansas University—home of the Muleriders. Andrea's father, Ron, was also an alumnus of SAU, and in memory of "Grannie Annie" (Andrea's mother) he had recently given the school a new mule, named "Molly Ann," to be its mascot. Andrea's entire family made the trip to Magnolia, Arkansas, where we all led Molly Ann onto the field to hand her off to the Mulerider at the halftime of a football game. On the way home, we talked about mules and what remarkable animals they are. I remember waxing poetic with some of our teenage children about the importance of the mule in American literature and culture ("forty acres and a mule"). And I had long harbored romantic, pastoral, and completely unrealistic illusions about mules and farms. Was there any great Southern literature that didn't have a mule in it? Almost everyone in Port William has a team of mules, and the passing from the mule to the tractor was the beginning of the end for the family farm. We were primed to be taken in by a mule.

So it shouldn't have been entirely surprising when, on a cold, rainy Saturday morning in February, Andrew came running into our bedroom saying, "Grannie Annie left us a mule on Craigslist!" There was the listing: "Molly Mule for sale. 7 months old. $100." Andrea was still skeptical, but Andrew and I drove out to a ramshackle farm outside of town to take a look at her. She was small and covered in mud, but we could tell she would be beautiful. She was already halterbroken, so the owners led her around the small, muddy pen she shared with assorted goats and chickens and a calf. Andrew's reasons for wanting her were almost as unrealistic as my own.

I called Andrea and let Andrew tell her that we thought this was a great deal. She said okay. We went into a nearby feedstore and asked if there was anyone around who had a trailer who would transport a mule for us. Eddie did some part-time work for the store, had a trailer, and said he could do it tomorrow. I offered him fifty dollars. Nobody asked us what the hell were we getting a mule for. The next day Andrea, Andrew, Eddie, and I all showed up to claim Miss Molly, our new mule.

Molly presented new challenges for us. We had not yet really attended to our fences, and now they had to be shorn up in a hurry. And for the first time, we would need to go to the farm (almost) every day to feed and check on her. Our friends Betty and Buster raise miniature donkeys, and they came out and taught us how to clean her hooves and use a rasp. Buster showed us how to pick up and hold her feet, stressing the importance of doing this regularly to desensitize her.

At first it was easy. She was like a big dog. On the second or third night we had her, she got out of our poorly reconstructed fence and wandered several hundred yards away from our barn into the surrounding pasture. When we got there, she came bounding up to us. Andrew could lead her all around the farm. She would nuzzle him, and she loved vanilla wafers.

Having a mule also means joining the community of "mule people." These are not people who act like mules but people who love them. Mule people will be quick to explain to you why mules are superior to horses. Though not as fast as horses, they are usually stronger, smarter, and much more sure-footed. This is why mules are used on trail and canyon rides. "Stubborn as a mule" doesn't refer to attitude—well, not entirely—it refers to the mule's intelligence and its unwillingness to put itself in danger. And mules are much easier to care for than horses. While a horse will eat itself to death, a mule will not. More than once we've had someone quote Harry Truman's line to us about why the mule was his favorite animal: "They know when to stop working and when to stop eating."

We were becoming mule people. I dreamed of the day when I would fully break her and hitch her up to a small wagon. On numerous occasions I proudly proclaimed to our guests that my goal was to be able to have her pull a cart by the time we had grandchildren. This would be our new thing. Children and grandchildren would come to our farm and go for rides in a one-mule open sleigh, or rather, wagon.

We bought books about mules and "the horse-powered farm." We went to a pulling demonstration at a fair. I read countless articles on the internet and in various magazines. I even bought a whole box of back issues of *The Draft Horse Journal* for a song off of eBay. I was (and am) in love with mules.

But our Molly was becoming difficult to manage. She grew and grew. She was strong and very smart. I kept telling myself that I was going to break her, but the truth is I never really put in the time. And as she got larger, she became ornery. She kicked Andrew and me on various occasions, and once she kicked Andrea's father so hard that it knocked him to the ground, probably fracturing a rib. After that Andrea put her on notice. She was going to shape up or she would be shipped out. Soon I was the only person in our family who could go into the pasture or barn with her. And once in the barn, she wanted only two things from me, either food or brushing. I began to have to pour her feed over the wall to keep from getting knocked over when I brought the feed bucket. And if she had already eaten, she would forcibly put her head under my hand for me to pet or brush her. This sounds cute, but it gets old in a hurry, especially if I'm trying to do something else like clean out her pen or trough.

We were told that she had to learn who was boss, and that a big part of this was who was in charge of the physical space. If she could make me move, then she was in charge. So I was supposed to get right up next to her, pushing her out of her space to show her who was boss, all the while her HUGE head is following every movement of my hands trying to get me to brush her. I almost always gave in. Most days I got stepped on. I regularly got pushed into the steel bars of the cattle shoot. And it was becoming increasingly clear to all of us who the real boss was.

The embarrassing thing for me was that I was realizing that I liked the *idea* of a mule more than the mule itself. For years I had made jokes about my students who like the idea of studying philosophy more than actually studying philosophy itself. Philosophy involves reading hard books and thinking through difficult problems. I would say that what those students really like is to sit in coffee houses and say deep nothings. They like the dark turtlenecks and imported cigarettes more than the butt-burning work of actually reading Kant's *Critique of Pure Reason*. And here I was, the same, only with mules. "Physician, heal thyself."

In time we realized that we needed to find a good home for Molly. We had read that while the average lifespan of a horse is twenty-five to thirty years, the average length of time a horse is owned is only four to five years. This is why there were so many horses for sale on the internet. And mules can live to be forty. We weren't going to sell her to someone who would do goodness knows what to her.

We found a ranch a few hours away that trained draft horses. The ranch owner, Janet, was a little incredulous when we called and said that we had a two-year-old mule that we wanted to *give* to her. She agreed to come take a look at her and then, maybe, take her. Once here Janet wanted her. When Janet put her halter on Molly and began to lead her to her own trailer, Molly began to put up a fight. But she was no match for Janet. Janet took hold of the lead and said, "Honey, the bitch is here. I'm the boss now." She loaded Molly, renamed her "Dolly," and took her away.

In the last couple of years, Janet has faithfully sent us pictures and updates of Dolly's progress. It's remarkable. It just took a trained hand who knew what she was doing and would put in the time to train her. Janet takes Dolly, fully trained now, on trail rides and uses her in a variety of ways. She's won a slew of ribbons and awards. Over and over again, Janet has remarked to us how smart Dolly is. This is no surprise to us. She was always smarter than I was.

SEPTIC MATTERS

We bought this little farm three years ago. Most of our attention has been focused on renovating the old farmhouse. We've had it leveled, added heat and air, removed and added walls, and made dozens of other renovations. We also put in two new toilets, and we've never had any trouble with the septic tank.

But of course we weren't living here full-time until a month ago. A couple of weeks ago, we started having some trouble with the toilets. Sometimes they wouldn't flush completely, but they weren't entirely stopped up either. They would slowly drain. A plumber couldn't find a problem. We hadn't had the septic system pumped out and figured that might be the problem. Solomon said, "Moore, you DEFINITELY need to get that tank pumped out."

So we had a pumping company come. They found the septic tank and emptied it. But they said the only thing that was running into the tank was gray water from the dishwasher and washing machine. Okay, so there must be another tank somewhere. But where? Initially we were afraid that it was UNDER the new deck that we just added on to the back of the house. If only...

Now we're down to only one working toilet, and we get another plumber to come with a line camera. Andrea met with him this morning while I was in the seminar. He discovered that there IS NO second tank. We had always known that the garage had been added later. Most of the house is pier and beam, but the garage sits on a slab. It turns out that the old farmer had dug a big hole behind the house that the toilets emptied into. When he added the garage, he built it ON TOP of the hole and just paved over it. The camera showed that it's got a gravel bottom and brick walls and contains decades of shit.

I spent my afternoon talking with septic people and setting up appointments for bids. It appears that in addition to putting in an all-new septic tank and replumbing the toilet lines, we are going to have to have the garage floor jackhammered, the shit pumped out, the hole filled up, and the garage repaved (and whatever else we can find to spend money on). Maybe once the experts get here, they will give us some better options, but I'm not hopeful . . .

ORPHAN CHICKS

Baby chicks are some of the great joys of the farm. We buy new chicks every spring. Tractor Supply calls it "Chick Days," and all of the local feedstores have an ample supply of new chicks in February and March. I love stopping by these stores on the way home to see if they have any unusual breeds (or older ones that they've discounted to make way for the new ones). And of course there are a fair number of chicks that are hatched here on the farm. We have an incubator and a brooder, and we've had some success with that. But the really exciting ones are the chicks that we simply discover in the barn or in a corral somewhere.

Sometimes a hen will have a secret nest that we don't know about, and she'll hatch her own brood. Sometimes a hen will lay an egg on a duck nest, the mama duck will sit on the nest, and lo and behold, there will be a baby chick amongst the ducklings.

One day this fall, Andrea discovered five new chicks following their mama around the cow barn and pen. For several days we enjoyed watching these little guys scratching and chirping and somehow managing to avoid being stepped on by the cattle. Mama was always on the lookout, and she wouldn't let us get too close to them.

One day when I was cleaning out the cattle water troughs, I heard the sound of one of the chicks chirping, but it didn't sound right. Looking around, I discovered that one of the chicks had fallen into a water bucket, and it was barely able to keep its head above water. Retrieving it, I tried drying it off by blowing on it, and then I took it to its mama and siblings. But Mama would have none of this. She did not receive the prodigal. Pecking and scratching it, she managed to toss it a couple of feet away, and she would not let it get any closer. Since it was probably my bad breath that made Mama reject the poor thing, it was now up to me to save it (again). I took it and put it in the brooder, and Andrew stole one of its siblings (much to Mama's dismay) for a companion.

For the next two or three weeks, we compared the progress of the two groups. Initially the brooder chicks on feed grew faster, but very quickly nature's own caught up and took the lead. But nature was dangerous as well. One day I found another chick in the water bucket, and this one was already dead. And about this time we had had a single little Bantam chick that had been hatched in a different barn, and now it was gone too.

At about a month old, when the chicks outgrew the brooder, we put them outside in the barn where we stack the hay. They began to have some interaction with all of the rest of the fowl who were coming and going in that part of the barnyard—ducks, geese, guineas, and lots of chickens. But these two little guys were orphans. They stuck to themselves, were always close together, and stayed near the house.

One morning we discovered that they had gotten locked in the garage overnight and had perched on one of Andrea's decorative wreaths. On another occasion they managed to make it on the screened-in front porch. When we were taking the two inside dogs out, they got a whiff of the chicks and chased them around the porch and then back *inside* the house. There was much commotion as we tried to catch the dogs and save the chicks, who had scampered all the way to one of the back bedrooms.

By the time Christmas rolled around, they were a couple of months old and seemed to be venturing farther away from the house. The hay barn and one of the chicken coops are not far from the dog yard, where the two Australian Shepherds live. The older ducks and chickens have learned to stay clear of the dog yard, and some bold roosters even seem to enjoy strutting in safety just on the other side of the fence. (The poor guineas never seem to learn anything, but flying over the fence into the dog yard is a mistake they only make once.)

We were only gone for one night over Christmas, but when we returned, we didn't initially see the orphan chicks anywhere. And then I found one—or at least what was left of it—in the dog yard. Just learning to fly, it had apparently taken flight only to land in the one place from which there is no return. Punishing the dogs at this point was pointless. Most people who raise backyard chickens know that the family dog is the most serious predator most chickens will ever encounter.

I didn't discover number two for a couple of days. I found it dead, lying under a wheelbarrow near the hay. There was no sign that it had been attacked. My sentimental side is inclined to think that without its companion it just went and hid and then died, but it could have died for numerous reasons: starved to death, stepped on by a goose or turkey, or given the unseasonably warm December weather, it could have been bitten by a snake. There's no way to know.

Mama Hen on the other side of the barnyard also lost another one. I never found its body, but there was only one remaining chick with Mama. So of the six natural chicks born this fall (the five plus the one Bantam), only one made it to the New Year. Survival of the fittest indeed.

SILKY SMOOTH'S BIG ADVENTURE

SATURDAY

Andrew's first heifer, Bessie, was not able to breed. We tried for almost two years using both AI and nature's way with no luck. The rancher from whom we bought her very generously offered to take her back and give Andrew another one. The one that he gave us, "Silky Smooth," is a beautiful ten-month-old red heifer, but she hasn't been broken. So for the last two weeks, we've been breaking her, and she has been showing remarkable progress. She stays tied up almost all of the time (pulling against the pole and learning that she can't win), and then we lead her to water and then back to feed. She learns to follow the lead because she gets to go to water and feed. All has been going very well. In the last several days, we have been able to walk her in large loops in an adjacent pen, and she has been responding very well.

Well, on Saturday it all went south. We had Shaun and Emily Anne here for the football game, and Shaun's parents even came out for lunch. We were walking them around the farm, and we got over to the open-air barn where Silky was tied up. I made a huge mistake, and it was all because of pride. I wanted to show our guests how well she was doing. I untied her, I gave Andrew the line to lead her to the water, and for a moment, everything was fine. But having more people than usual in the barn (even behind the rails) must have spooked her.

And all of a sudden, this over-eight-hundred-pound heifer took off. She ripped the line out of Andrew's hand, broke through three lines of barbed wire in the adjacent pen, and ran into the front corral. Further spooked by our donkey, she took off toward the road, breaking through another barbed-wire fence in the front, and ran down the road at a full sprint. Andrew never missed a beat. His hand bleeding from the rope burn, he jumped the broken fence, ran through the corral, hurdled the front fence, and headed down the road after his cow.

The cow eventually found her way into a large plowed field about a mile from our house. Shaun and his father hopped in their car to follow Andrew. I hooked up the trailer and headed after them. Silky had made her way to the far back of this huge field (maybe 150 acres?). Andrew was out there, but the cow was a long way away. I parked in the driveway of a small house set back from the road, climbed through another barbed-wire fence, and headed toward Andrew. The plowed field is hard to walk in. I could see Andrew making his way toward the cow (who still had her halter attached). He was getting close, and Silky was standing still. I thought, "Okay, he's going to get the lead line." And then Silky charged Andrew. I was probably two hundred yards away, and I saw her lower her head and hit Andrew squarely in the chest, lifting him in the air and knocking him to the ground. I'm running as fast as I can, but I can't get much traction in the soft dirt. When I finally reach him, I realize that he's scraped up and his chest hurts, but he's okay. Thanks be to God. But the cow is now a long way away.

Back home the rest of the family has scrambled to put the other animals away so that they won't also escape through the new gaps in the fences. Hannah and her friends corral the donkey and move her to an inside pen with the sheep. The Jersey heifer, Annabelle, has seen the gap and made it into the front field that has been closed off to her all summer. Annabelle is Andrea's cow. She bottle-fed her three times a day for the first three months of her life, and now at five hundred pounds, Annabelle bounds after Andrea back to the safety of the barn.

By the time Andrew and I get back to the side of the pasture where we've parked the truck and trailer, Andrea is there. She's also been on the phone with some neighbors who raise cattle, and they're offering advice. She's brought a bucket of feed for the cow and Gatorade for Andrew and me. The neighbors say the heifer's got to settle down. See if she'll follow the feed bucket. She seems calmer, but she's staying well clear of us. Our neighbor Lyndon shows up in his four-wheel drive truck, and the next couple of hours are spent trying to herd the cow to a pen on the far side—first with the truck and then with four-wheelers. No luck. She simply won't drive. At one point they get her fairly close to where I am, and I believe that I can get that line. I know her pretty well, I've been working a lot with her, and I believe that she'll trust me. Lyndon thinks it's worth a shot but he tells me to make a little circle if she charges. As I get close, she begins to scrape the ground with her front hoof, just like in those bullfighting movies. A wiser man would have taken this as a sign.

I don't think I've mentioned that we're dressed for the Baylor game. This was, after all, supposed to be a pregame party. Hannah and her roommates were coming out, and of course the Kuonis were there. Andrea had spent all morning baking fresh rolls for the sandwiches, which we would garnish with our own fresh pickles, peppers, and tomatoes. There was spicy mustard to go with an assortment of deli meats and cheeses, and she had made hot brownies for dessert. All of this was lavishly spread out and untouched in our kitchen. And I was staring down a spooked heifer in a plowed field, wearing a bright yellow Baylor shirt and starched white cotton shorts with slip-on shoes that look for all the world like slippers.

Andrea had just left to go back and check on the guests. If she had still been there, I know she would have told me to get away. But "making a little circle" seemed like a simple enough maneuver. You step to one side and let the cow run past. And when the inevitable occurred, that's what I tried to. But the ground was so soft that when I moved, I just fell. Eight hundred pounds of USDA Prime beef brushed beside me as my starched white shorts met the freshly plowed Texas dirt. She missed me, but now where was she? As I scrambled to my feet, knees bleeding and heart racing, I turned to find her fifteen feet away and doing the hoof thing again. It turned out I can move fast—this time back behind a tree.

After thirty more minutes of trying to drive her with the four-wheelers, we decide that we need to let her calm down overnight. Andrew's four-wheeler won't start, and now it looks like we're going to have to leave it there as well. Throughout this ordeal we've had various spectators show up: the young couple living in the house whose yard we've invaded; Shaun's parents, Shaun and Emily Anne; Lyndon's wife, Tracy; and her father, who owns the adjacent land. Most of us are decked out in our Baylor green and gold.

Back home we're tempted to skip the game, but Samuel is going to be running in the Baylor Line for the first time. I go back to the pasture and try and close off the two openings to that field to keep her in there. My phone starts going off, but I'm not paying attention to it. Andrea calls and says, "Are you reading your texts?" No, I'm not reading the texts. I'm trying to build a barrier that will keep the cow from coming through the very opening she has been avoiding all afternoon." "Read the texts!" Samuel—who hates group texts with a passion—has sent a message to us and all of his siblings saying, "I'm about to join the Baylor Line. I'm going to fulfill my birthright." Okay. We're going.

Andrea and I race to the stadium to make it for the pregame entrance of the Line. We make it in plenty of time to find that Samuel has inched his way almost to the very front of the Line. He and his siblings are all texting—even Benjamin, who is at the Notre Dame–Virginia game in Charlottesville. Dad and Skip have joined us at our seats. They are wearing matching Columbia green-checkered Baylor shirts and khaki shorts. Dad may have trouble remembering some things, but not "That Good Old Baylor Line." Bear claw aloft, he sings with gusto every word. When the time finally comes, Samuel sprints to the front, racing the hundred yards to the other goal line. It was worth it to get here.

Andrea and I stay for just the first quarter of Baylor's game against Lamar. It's a sloppy game with Lamar taking the lead, 14–13 just after we left. It's dark by the time that we get home, and Andrew and I drive to the pasture but cannot find Silky in the dark. The Bears win 66–31 despite four turnovers and lots of mistakes. The first day was over. Silky 1–Moores 0.

SUNDAY

"And he answered them saying, 'Which of you will have an ass or an ox fallen into a ditch and will not straightway pull him out on the Sabbath day?'" (Luke 14:5)

Sunday morning was the most beautiful morning we've had in Central Texas in months. The cool front had come through, and the temperature had fallen into the 50s for the first time since spring. Andrew and I went back to the pasture. The first order of business was to get the four-wheeler running again. Tightening the battery wires and adding more gasoline got the job done. With both of us on board, we went in search of Silky. We found her lying down in the high grass in the far corner. I approached her slowly, and she got up and trotted off. We let her be.

Back at the house, Andrea went to church to teach Sunday school and work extended session. Andrew, Shaun, and I set about the business of repairing the fences and the pen that Silky had broken the day before. After lunch we got reinforcements. Lyndon and Tracy Love run cattle on a couple of ranches with more than a thousand acres. They showed up with a friend of theirs and saddled horses, ready to put Silky in her place. While the three of them set off across the plowed field, Andrew and Shaun put up some cattle panels to close off an opening in the pen to which they were going to drive the cow. It would only be a matter of time now. I waded through the high grass to try to open the long-rusted gate to the pen. Struggling with the gate, I disturbed a wasp nest and promptly got stung. After I found two more nests, Andrea and Emily Anne returned to the house to get the wasp spray.

The far corner of the pasture is probably half a mile from the pen where we're watching the action from the bed of our truck. Today the gallery includes Emily Anne, Shaun, Andrew, the Loves' daughter Sally, Andrea, and me. We are no longer wearing green and gold. Twelve-year-old Sally, already a champion barrel racer, is teaching our family how to rope a five-gallon bucket. It's very hard to see what's happening across the field. The three horses have driven the heifer out into the opening, but they don't seem to be heading our direction. They circle, they drive, and then they retreat. For more than an hour, three experienced cattlemen fight with this cow. The cow will not drive. Rather, she charges the horses and their riders. After an hour they leave her be and make the long ride back to our perch. One proclaims, "She's the meanest cow I've ever seen." Another replies, "I think you're probably gonna need dogs for this one."

I call John, the ranch manager from whom we got her, seeking advice and hoping that he will bring his dogs and just come solve the problem. It's a Sunday afternoon, and his cattle hands have the day off. He'll have to call me back. An hour later he calls to suggest that we leave a feed trail ("kind of like Hansel and Gretel") all the way to a tub of water in the pen. "If that don't work, call me in the morning." Skeptically we comply and hope for the best. Day two was over. Silky 2–Moores 0.

MONDAY

Monday morning Andrew and I rose early to go see if by chance the "Hansel and Gretel" strategy had worked. It had not. A missing heifer seemed to Andrew the perfect excuse to get out of school (and football practice), but we weren't buying. Amidst his passionate protest, Andrea took him to school and said we'd let him know if anything happened.

At about eight thirty, Andrea and I decided that we needed to go put eyes on her. We drove the truck into the pasture in search of her. I was a bit anxious about this, and we kept our "two-wheel drive" truck on the strip of grass surrounding the large expanse of plowed field. Getting the truck stuck in the mud could only add insult to injury, and we feared the neighbors had already begun to talk.

Silky, however, was not to be found. For the last two days, she had been retreating to some high grass in the far corner, but she wasn't there now. I walked the length of the back fence line (which couldn't be driven on) while Andrea took the truck back the way we came. No Silky. And there was no sign of her: no cow patties, no crushed grasses, no broken fences. There were a couple of tanks on the other side of the fence, and I knew that she would have been without water now for a couple of days. But she was gone.

Now for the first time, I really begin to feel sick. Have we really lost her? We begin to make calls and send texts to neighboring farmers and ranchers to keep an eye out for her. With nothing more to do, I head to campus and to work. For some time I had been planning on going to Austin to hear Robert Wilken give an address that evening. I almost give up on this, but since there is nothing more to be done with Silky, I head south with friends Doug and Coleman.

Andrea calls en route. After picking up Andrew, they had gone in search of Silky and found a red Beefmaster with a white ear tag (but no halter) in one of David Miller's herds, more than a mile from where we'd last seen her. She was quite a distance away in the pasture. Could this be Silky? Yes, it was. What was lost was now found. Silky 2–Moores 1.

IN THE DAYS THAT FOLLOWED

During the next week, Lyndon had confirmed that now it would just be a matter of moving that herd and closing gates behind them until Silky could be moved into a loading pen. All of this had happened with little fanfare and no involvement from the Moores. Andrea, Andrew, and I checked on her off and on during the week.

Early Saturday morning Lyndon joined Andrew and me, and we took our trailer to pick her up. Having the right facilities and tools makes all the difference. Loading her in the trailer now took less than five minutes. She was mad as hell but in the trailer. Sometimes *techne* is superior to *phronesis*.

We took her back to Axtell and to the ranch from which we got her. The ranch manager (and his dogs) met us, and we unloaded Silky into a new pen. He would work with her and try to break her enough so that we could manage her. He offered to buy her from us, but we declined. If he wants her, that means that she's worth keeping. So now the real cowboys will work on her for awhile, and we'll get her back in a couple of weeks. Maybe one day she really will be silky smooth. Until then it's probably a draw.

Silky 2–Moores 2. To be continued.

TO A HARE, FROM A LOUSE

"Wee, sleekit, cowrin, tim'rous beastie,/O, what a panic's in thy breastie!"[1] This is the beginning of one of Robert Burns's best-loved poems, "To a Mouse." In the poem Burns reflects on the plight of a poor mouse whose burrow he accidentally destroys while plowing one blustery November day in 1785. (Why Burns is plowing in November in Scotland is less clear.) Burns laments that he has left the mouse without a home as winter is approaching and concludes with the famous lines, "The best-laid schemes o' mice an' men/Gang aft agley,/An' lea'e us nought but grief an' pain,/For promis'd joy!" And it is of course from these lines that John Steinbeck draws the title of his own immortal *Of Mice and Men*.

But it was another passage from that poem that came to mind as I was mowing one of our pastures the other day. Early in the poem Burns says:

> I'm truly sorry Man's dominion,
> Has broken Nature's social union,
> An' justifies that ill opinion,
> Which makes thee startle,
> At me, thy poor, earth-born companion,
> An' fellow-mortal!

I often encounter small creatures when I'm mowing in the pasture. Like Burns I mostly see field mice that scatter in front of the mower, but I also encounter rats and rabbits and the occasional snake. On several occasions I have stumbled upon large clutches of eggs laid in the middle of the high pasture grass. These are usually from our guinea fowl. Sometimes we have collected these eggs and put them in the incubator, but we haven't had much luck hatching guineas. One never knows how old the eggs are, and getting eighteen or twenty eggs all the way from a pasture to the house without dropping a few is no small feat. It's certainly an inconvenience to stop mowing and tend to the eggs.

But on this occasion, it was not a clutch of eggs that I disturbed while mowing. It was a rabbit. She was in the high grass near where the aerator sprinkler spreads the water from the Cottage's septic system. As the mower neared, I saw her hop and move through the high grass. When she didn't run away, I figured that her burrow must be near. She stayed in the high grass, moving as I moved, but she didn't scatter. I continued mowing, but I was slow and careful, peering as best I could into the grass I was determined to eliminate. Once Andrea had encountered a rabbit's nest in the other front pasture, and she had dutifully avoided it, leaving an obnoxious patch of weeds all summer in the middle of the donkey's pasture. I knew my duty. And I remembered Burns.

I was almost finished mowing this pasture, and when I came to the end of my row, I noticed that something was sticking out of my front left tire. There was a piece of baling wire, three inches long, protruding out of the tire. I got off of the mower to investigate. There was red slime around the edges of the wire (the "red stuff" is the cheaper version of the "green stuff" used to seal tires). I knew that if I pulled the wire out, there was a chance that the hole would be too large, and the tire would immediately begin to deflate. But it was possible that the wire wasn't in too far, and when I pulled it out, the red stuff would do its job and seal the hole for the time being. If I left it where it was, it would almost certainly penetrate farther and leave me with a flat tire in the middle of the pasture. To pull or not to pull, that was the question.

Flat tires in the middle of the pasture are always complicated endeavors. Not only is there the obvious inconvenience that any flat tire produces, but it's made more difficult in a variety of ways. Front and rear tractor tires are not the same size, and one might not have a spare. That means either repairing the tire or a trip to the feedstore—the loss of an hour and a half either way. Even if I do have a spare, it might be the wrong size. If one wheel or tire is ever so slightly different, it will leave all of your rows uneven—the grass will be taller on one side of every row. This isn't a bad problem in the pasture, but it's big deal if you're going to be using the same mower on the lawns around the houses and barns. Moreover, changing the tire is not always an easy task. First you've got to walk all the way back to the barn where the jack is kept, and then you have to carry not only a heavy jack and the spare tire back into the pasture, but you probably also need to bring some boards or bricks to put under the jack to keep it from sinking into the soft dirt once you begin to raise the machine. But small lawn tractors are not that high off of the ground, so it's entirely possible that there will not be enough space to put both your board and your jack underneath the now disabled mower. And of course you're in high grass with no shade, which makes everything more difficult, and even if you're successful, you'll have to cart all of this equipment back to the barn before you can resume the mowing that should have been a simple and enjoyable task.

All of this is in the back of my mind when I make the momentous decision to pull the wire out and take my chances. A small stream of air starts coming out, and my only thought is that perhaps I can finish these last two rows and get the mower to a place where it will be easy to change the tire. Now the clock is ticking, and I've got to go fast.

I speed down the row, knowing that with every second air is coming out of my tire. When I make the turn at the end of the row, I catch the sight of something moving in my peripheral vision. Oh no. I make a broad circle and turn around on grass that has already been mowed. There is a very small rabbit that looks at me and then disappears into the cut grass and weeds. I begin to scour the surrounding area, and then I see it. In my haste I have run over the rabbit's burrow, and I have done worse than merely destroy her home. There are three dead tiny baby rabbits nestled together, and I have killed them.

I turn off the mower, no longer worried about the tire. I look around the pasture. Neither mama nor the other baby is anywhere to be found. When I examine the nest, I discover that there is very little blood, and at first glance, the three tiny rabbits seem to be sleeping. Only they are not.

There is nothing to be done. I am a little sick in my stomach, and I am angry at myself for my carelessness and stupidity. I try to console myself by thinking that this is just how nature is. Survival of the fittest and all that. But of course it wasn't nature that killed them. I did it. And I did it because I was in a hurry and I wasn't paying attention and I didn't want to be inconvenienced. When I return to the mower, I discover that the red stuff has apparently sealed the tire, and it's no longer deflating. I finish my row and make the long, slow drive back to the barn.

By now I'm thinking of Burns and how "man's dominion has broken nature's social union." Is there really a social union with the rabbits? Was Old Major right after all when he told the other animals, "Never listen when they tell you that Man and the animals have a common interest, that the prosperity of one is the prosperity of others. It is all lies. Man serves the interests of no creature except himself." When it comes to rabbits, aren't I really more like Mr. MacGregor chasing them out of the garden and away?

Well, sometimes. There is the rabbit we've named "Peter" who lives under the deck at the Cottage. He greets us most mornings and evenings, and we leave a little lettuce for him at the edge of the deck. There are the birds that make their homes in our many birdhouses and of course the peaceable kingdom of the barnyard. It's a union of sorts.

Burns also wrote a companion poem entitled "To a Louse." In that poem he tells of sitting behind a pious, sophisticated, well-dressed woman in church and observing the lice that creep through her hair and bonnet. Burns suggests that the louse should be on a beggar's head or even a small boy's coat. In mock rebuke he scolds the louse for his impudence at climbing all the way to the top of this fine, devout woman's bonnet, "Your impudence protects you sairly." And to "Jenny," the pious, unknowing woman, "You little ken what cursed speed/The blastie's makin!"

But on this day, I am the louse. It's my "impudence" and "cursed speed" that has made me break what little union I had with my fellow mortals. And it's all the worse because of my own self-satisfaction. I pride myself on going slowly and being careful, smugly quoting

Simone Weil on attention and Gene Logsdon about "living at Nature's pace." And all the while I'm recklessly speeding through the pasture because I don't want to be inconvenienced.

> O wad some power the giftie gie us
> To see oursels as ithers see us!
> It wad frae monie a blunder free us.

FARMERS, CHRISTIANS, AND INTELLECTUALS

CULTIVATING HUMILITY AND HOPE

As a professional philosopher who is also a practicing Christian, I'm used to the stares of incredulity that often come from the cultured despisers of our age. Had I not heard that religion was mere superstition? Was I unfamiliar with Freud's understanding of illusion and wish fulfillment, or Marx's opiate, or Nietzsche's death of God? Did I not know that smart people outgrow God? Yes, I've heard. If the culture of American higher education is overwhelmingly secular, the world of professional philosophy is even more so. Though there are counterexamples at every hand, it is true that the most commonly held belief by faculty in higher education today is the notion that "smart people outgrow God." From the most obscure community college to the most prominent research university, religious belief is often treated like superstition and understood antecedently to be a cultural phenomenon that will wither away under the inevitable march of social evolution. Robert Bellah chronicled thirty years ago how "leaving home" and "leaving religion" were correlated in the minds of most academics, and this sociological perspective has been confirmed time and again.[1] In the modern university, if you want to know how faith and learning are related, most of higher education will insist that the more learning you get, the less faith you need.

But the skepticism about Christians and intellectuals pales in comparison to the assumptions about intellectuals and farmers. The rural world and all its inhabitants are regularly assumed to be one large, indistinguishable mass of ignorance and unsophistication. If ever there were a class of people regularly derided for their lack of intellectual rigor or cultural vitality, it is the rural populace. There is almost no end to the epithets: hicks, yokels, rednecks, hillbillies, rubes, country bumpkins, hayseeds—the list goes on and on. Where were Jefferson's agrarian intellectuals? Where were the citizen farmers who read great books by lamplight and quoted Wordsworth behind the plow? Was all this merely the romantic idealizations of a lost age?

So it is easy to imagine that a so-called intellectual who aspires to be both a faithful Christian and a successful farmer, might just turn out to be none of the above—or be consumed by consoling delusions. That's always a possibility. However, I think that this is not the case. It seems to me that these three identifications—Christian, farmer, and intellectual—are actually mutually affirming of each other, and they function as correctives to some of the worst impulses and temptations given to all three. To understand how this might be so (and to understand the dangers that reside here), let us start with a look at our terms. Aristotle said that any good investigation begins with what we know best and moves toward what we know least—or not at all.

"Farmer" is perhaps the least controversial or ambiguous of these terms. A farmer is someone who is engaged in agriculture—one who cultivates the *agros*, that is, the fields, soils, crops, and animals that are necessary for human life and food. Farmers have always transcended the "Old Macdonald" stereotype, and these days "farmer" might refer to everything from the agribusinessman running thousands of acres of fertilized soybeans to the urban hipster who grows kale and Swiss chard in raised beds on a Brooklyn rooftop and sells them at one of the thousands of new farmers' markets. And of course there is just about everything in between.

An "intellectual" is more than just a smart person. An intellectual is someone who has worked to develop his or her intellect, his or her mind. This means the development of understanding and perception. *Intellectus* was the Latin word most often used to translate the Greek term *nous*—meaning both "mind" and that intellectual virtue that is the knowledge of first principles. There are many, many activities and vocations that require extraordinary intelligence but that, nevertheless, do not cultivate the life of the mind. Being smart does not make one an intellectual. An intellectual is someone who recognizes that the complexity of the world demands rigorous attention. The nature of the world, in both its beauty and its absurdity, is not a puzzle to be solved or a discrete question for which there is a particular, singular answer. There are of course very difficult questions that are sometimes solved by singular, elegant solutions. But the complex reality of the world is not one of them.

The complex reality of the world is more like a plot of land that requires careful, perpetual attention and husbandry. There is a circular nature to both the knowledge and the care of this plot of land. One cannot learn all there is to know about it without actively caring for it, and one cannot adequately care for it without careful attention to it and to all that affects it. So it is with the nature of the world. An intellectual is one who is ever in the process of gaining the competencies (both the knowledge and the skill) and the dispositions to attend to the complex character of the world.

So what does it mean to be a "Christian intellectual"? How does the adjective modify the noun? Well, a Christian is a follower of Jesus Christ and a member of the Body of Christ. It is someone who recognizes that the grace of God is utterly transformative for how one understands oneself, one's place in the world, and the nature and character of the world itself. Thus, when it comes to understanding the nature of the world, a Christian is someone who recognizes the transformative reality of the grace of God. This grace is no mere metaphysical addition to the world, a sixth sense like night vision or the ability to see color in a black-and-white world—seeing what others cannot. Yes, we do perceive a reality that is often unknown to or unappreciated by our non-Christian friends and colleagues, but the notion of "grace as technicolor" metaphor fails to do justice to the reality of grace. Grace is not just "there." It is the unexpected and undeserved reality that crashes in and saves us from ourselves and from our carefully laid plans for self-destruction. It is oxygen to the submerged swimmer, the cool drink of water to the field worker, and rain to the parched field itself. Christians recognize this grace as triune in nature, and this triune God is both the giver and the gift itself.[2]

In short, a Christian intellectual is one who is being transformed by the reality of grace while vigorously attending to the life of the mind and the complex reality of the world. Notice I did not say anything about the political culture wars. Being a Christian intellectual does not commit one to certain views on evolution, school prayer, abortion, pacifism, capital punishment, or—for crying out loud—Obamacare. Now, I believe that there are *certain* views on *some* of these subjects that would be difficult or even impossible for a Christian to hold—but that's not at all the same thing as suggesting that a Christian intellectual must hold such and such a view. When it comes to politics, we would do well to remember Max Weber's famous observation that "politics is a strong and slow boring of hard boards"—and that goes for Christians working in politics as well. In fact, if I'm right about Christian intellectuals recognizing the complex reality of the world, then we should expect Christian intellectuals to be less than satisfied with the cheap and easy bumper-sticker philosophies that populate the political left and right.

Once again, a Christian intellectual is one who is being transformed by the reality of grace while vigorously attending to the life of the mind and the complex reality of the world.

I

I think it should be obvious how being a farmer follows and contributes to this notion, but before turning to this question I should say a bit more about the complex reality of the world. To attend properly to the complex reality of the world, the Christian *must* cultivate two difficult virtues, humility and hope. They are not only essential virtues for all Christians but they are particularly important for those of us who aspire to cultivate our minds in the pursuit of truth. My students often find themselves vacillating between exuberant self-congratulation for their not insignificant accomplishments and fearful despair of finding themselves in the deep end of the ocean with no relief in sight. Humility and hope are often overlooked or misunderstood. Humility is not self-deprecation or low self-esteem, and hope is not optimism or consoling fantasy. Humility and hope are *theological virtues*. They have their origin and their *telos* (that is, their goal or their perfection) in the love of God and the community of the Body of Christ, which is the Church. And they are intimately related.

But before we can talk specifically about humility and hope, I need to say just a few words about the virtues themselves. It is important to remember what virtues are. If we rely on Aristotle and St. Thomas Aquinas, we know that virtues are dispositions of character that enable a thing to become what it was designed to be. Sharpness is a virtue of carving knives because carving knives are designed to slice, and flatness and dullness are virtues of butter knives because butter knives are designed to spread. We know whether anything (a knife, a pen, a movie, a friend, a sweater) is a "good one" if it has the characteristics that enable it to flourish at the task that is given to it. Is this wide-tipped Sharpie a good pen? It depends. Do I want to label a box to be shipped home or make notes in the margin of my Bible? The same goes for human beings. What are those characteristics (virtues) that human beings must have if they are to flourish at the task of "being human"? Aristotle and Thomas both believe that "flourishing" is just "being happy as a human being."

Moreover, Aristotle reminds us that virtues are mean states (that is, in-between—not nasty—states) that are found between states of deficiency and states of excess. Thus the virtue of courage is the mean between cowardice (the deficiency) and rashness or foolhardiness (the excess). In the case of the virtue of courage, we can only learn to identify the virtuous mean by living with courageous people and cultivating those habits that will help us follow their examples.

Before turning to humility and hope specifically, it is important to remember that virtues are not just "memorized"—though there is much to learn. They are "cultivated"—you might even say that they are grown. For the moment I am not going to worry about the difference between "infused" and "acquired" virtues, but the important thing to remember is that becoming virtuous is more like becoming a good carpenter than it is like studying hard and making an "A" on a test.

II

What is humility and how is it achieved? What are the false forms of humility that need to be guarded against and how do we know the difference? It is not only a difficult virtue to achieve, it is even hard to name and describe. It is of course the corrective to pride and arrogance, and we are inclined to associate it with self-effacement, and there is surely something to that. But humility is not a tendency to "put oneself down." It is not even a tendency to underestimate one's abilities or to think less of oneself than one should—and it is certainly not just to *appear* to think less of oneself. Humility must not be confused with pusillanimity, that smallness of soul that is one of the vices of magnanimity. In fact, Thomas Aquinas is very clear that humility does not conflict with magnanimity or greatness of soul. I need to be careful that I do not stray too far from the path just as we are trying to understand humility, but we can never forget that the virtues come on the group plan; we are not allowed to pick and choose the ones we like.

So what is magnanimity and how does "great soul-ness" not conflict with humility? To quote Josef Pieper, magnanimity is "the expansion of spirit toward great things; one who expects great things of himself and makes himself worthy of it is magnanimous." The magnanimous and the humble person neither complains or brags, and she shuns flattery and hypocrisy, all of which are unfitting to a great soul that knows its true worth and its calling.[3] Pusillanimity is just the smallness of soul that retreats from and refuses to rise to the challenges placed before it. This is not humility. To pusillanimity we should of course add self-abnegation and masochism. All of these are "false doubles" for humility.

Humility finds its beginnings in temperance or self-control. It is one of the few virtues that is explicitly both a moral and an intellectual virtue. We must be humble with respect to how we act, what we think of ourselves, and what we think we know about others and the world around us. In fact, moral humility makes intellectual humility possible, and vice versa.

Humility is thus a proper rendering of the self and of the self's relation to others and the world. But to call it a "proper rendering" does not really help us much, does it? Usain Bolt may accurately believe that he is (or was) the fastest human being ever, but none of us will think of him as an exemplar of humility. No, the proper rendering of oneself is to recognize first and foremost that the self is not the first thing. Put another way, the truth of the matter is that most of us need to get over ourselves. Whether focusing on our accomplishments or our inadequacies, most of us spend far too much time thinking about ourselves, and the first task of humility is to think about someone else.

I have learned a great deal about humility from the Anglo-Irish philosopher and novelist Iris Murdoch. In one of her novels (*The Nice and the Good*), one of her characters describes damnation as "one's ordinary everyday mode of consciousness [being] unremitting agonizing preoccupation with self." This same character describes happiness as "a matter of one's most ordinary everyday mode of consciousness being busy and lively and unconcerned with self." I think that's pretty good. In another place, Murdoch notes that most of our moral failings can be traced to "the fat, relentless ego."[4]

Most importantly, humility fosters what Murdoch called "a strong, agile realism." What does that mean? It means that humility makes it possible to see the world as it is. Seeing the world as it is cannot but help to cultivate humility because it turns our gaze away from ourselves and toward others and the world. Murdoch writes, "To know oneself *in the world* (as part of it, subject to it, connected with it) is to have the firmest grasp of the real. This is the humble 'sense of proportion' which Plato connects with virtue. Strong agile realism, which is of course not photographic naturalism, the non-sentimental, non-meanly-personal imaginative grasp of the subject-matter is something which can be recognised as value in all the arts, and it is this which gives that special unillusioned pleasure which is the liberating whiff of reality."[5]

Note the ways in which Murdoch here subverts our selfish intuitions. We might expect that the "firmest grasp of the real" would give us a humble sense of proportion. We do not expect that the "liberating whiff of reality" will give us "unillusioned pleasure." Most of us are antecedently inclined to associate humility with humiliation—and indeed we should—they come from the same root. Murdoch notes that this accurate self-knowledge is liberating and *pleasurable* beyond all our expectations. Such self-knowledge is humorously—but unsurprisingly—less and less focused on and absorbed by the self. G. K. Chesterton in *Orthodoxy* remarks, "How much larger your life would be if your self could become smaller in it."[6]

(Some readers might find it curious that I appeal to Murdoch's understanding of attending to reality in the midst of a *theological* reflection. Murdoch quite explicitly denied the existence of the traditional understanding of God. [Depending on which "tradition" we are talking about, I might as well.] But Murdoch's atheism is a curious phenomenon and addressing it satisfactorily is far beyond the scope of the current essay. She has been called "the most religious atheist of the twentieth century." For more on this, see "Iris Murdoch's Vexed Relationship to Christian Faith" in this volume.)

Returning to Murdoch, she connects humility to freedom and liberation. She describes true freedom "not as unimpeded movement but as something very much more like obedience."[7] This "obedience to reality as an exercise of love ... [is] surely ... the place where the concept of good lives." It lives in contrast to the selfishness of the fat relentless ego. Goodness is "the attempt to pierce the veil of selfish consciousness and join the world as it really is." Humility is this "virtuous consciousness."[8]

This virtuous consciousness is characteristic of good students and serious artists, and I think, successful farmers. In a famous interview that Murdoch did with Bryan Magee, she connects the humility of the artist with the recognition of reality. The serious artist "feels humility since he knows that it is far more detailed and wonderful and awful and amazing than anything which he can ever express. This 'other' is most readily called 'reality' or 'nature' or 'the world'...."[9] "The honesty and humility required of the student—not to pretend to know what one does not know—is the preparation for the honesty and humility of the scholar who does not even feel tempted to suppress the fact which damns his theory."[10] Farmers are even better prepared to understand this than scholars because the farmer is trying to feed his or her family, not prove a theory.

Humility—or humiliation—maybe the easiest crop for a novice farmer to cultivate. How have I embarrassed myself in front of other farmers? Let me count the ways. Learning how to back up a trailer into small spaces, digging and redigging post holes to try and get the fence line straight (and they never are), chasing loose livestock, overgrown yet underproductive gardens, and untold numbers of stupid questions asked to farmer friends and at the feedstore. All of this can be pretty humiliating, but an honest confrontation with the realities of farm life is also liberating in an odd sort of way.

But too much realism can also be pretty depressing, for both the scholar and the farmer. It pushes us toward either presumption, despair, or both. One of my favorite Wendell Berry characters, Wheeler Catlett, remarks, "A man who is depending on the truth to console him is sometimes in a hell of a fix."[11] And that brings us to hope.

III

Hope is not optimism. Neither is hope a fantasy that consoles against realities we fear will come to pass. True hope is a sure companion to humility. I know of no one better on hope than the German Catholic philosopher Josef Pieper. For Pieper hope only becomes a virtue by being a "theological virtue." According to Pieper, "Theological virtue is an ennobling of man's nature that entirely surpasses what he 'can be' of himself. Theological virtue is the steadfast orientation toward a fulfillment and a beatitude that are not 'owed' to natural man."[12] For hope this means that it is a "steadfast turning toward the true fulfillment of man's nature, that is, toward good, only when it has its source in the reality of grace in man and is directed toward supernatural happiness in God."[13]

The heart of Pieper's analysis of hope resides in his articulation of hope as being found in the mean between despair and presumption. Indeed, both despair and presumption are the forms of hopelessness that "destroy the pilgrim character of human existence." Pieper reminds us that despair is not a mood but rather an act of the intellect and a decision of the will, and as such it is sin. He notes that Peter Lombard counted despair among the sins against the Holy Spirit, and for Pieper despair "moves us into the vicinity" of the mystery of such sin because it "closes the door" on the reality that is the way of fulfillment brought about by the Holy Spirit. It is "by its very nature a denial of the way that leads to the forgiveness of sin."[14]

I want to pause here a moment and emphasize that Pieper draws the distinction between reality (as it is in itself) and the rejection of that reality by the self-imposed way of the self, especially as it has its roots in *acedia*, sometimes translated as sloth or spiritual sadness. It is important to recognize that at the heart of these questions about the nature of hope and its power to offer authentic encouragement and consolation is the question of how we understand the world to be. And this inevitably connects hope to humility and thus to contemplation and ultimately to the flourishing that is happiness.

Pieper's analysis of the root of despair, that is, its relation to *acedia*, is among his most moving and most persuasive, and, regrettably, for many of us—especially teachers and students—most damning. As is often noted, the traditional rendering of *acedia* as sloth fails in the extreme to do justice to the concept. In classical theology it is a kind of *sadness*. "This sadness because of the God-given ennobling of human nature causes inactivity, depression, discouragement." I quote Pieper here at length:

> The opposite of *acedia* is not industry and diligence but magnanimity and that joy which is the fruit of the supernatural love of God. Not only can acedia and ordinary diligence exist very well together; it is even true that the senselessly exaggerated workaholism of our age is directly traceable to *acedia*, which is a basic characteristic of the spiritual countenance of precisely this age in which we live.

According to Pieper this sadness is a "lack of magnanimity; it lacks the courage for the great things that are proper to the nature of the

Christian." "One who is trapped in *acedia* has neither the courage nor the will to be as great as he really is." It is a kind of "perverted humility [which] will not accept supernatural goods because they are, by their very nature, linked to a claim on him who receives them." It is the hopelessness borne of despair.[15]

As despair shows the deficiency of hope, so presumption shows the vicious excess. Presumption is a "perverse anticipation of fulfillment," "an attitude of mind that fails to accept the reality of the futurity and 'arduousness' that characterize eternal life." Presumption leads one to deny the essential pilgrim character of human existence while seeking, and even believing that one has achieved, a security "so exaggerated that it exceeds the bounds of reality." If despair has its roots in a failure of magnanimity, then presumption is grounded in a lack of humility that places its confidence in the will and in the misdirected use of the intellect.[16]

On the farm the misdirected use of the intellect is often tied to an exaggerated confidence in technology to solve our problems. And these problems usually take the form of attempting to achieve security by "beating nature," whether through the use of dangerous chemicals, expensive machinery, or by the mere presumption that we can force nature to comply to our will. It should be no surprise that so many modern farmers find themselves strangled by presumption on the one hand and despair on the other. But the farmer's situation is no different from the urban businesswoman who is at the mercy of the companies who make and control apps on her cell phone or the single parent struggling to find enough time to make a living wage and still parent her children. Our consumer culture repeatedly promises us that a technological (or financial) remedy will solve our problems, and we despair when we finally recognize that even if their technological solution works (not a given), it will require ever more technology (and money to pay for it).

Presumption falsely believes that fear can be banished and fails to recognize how self-deceptive and unnatural the posture of "fearlessness" is. We live in a culture that thinks that "fearlessness" is, or would be, a good thing, but IT IS NOT. "Fearlessness" is a dangerous presumption against which we must guard with both temperance and perseverance. At the conclusion to his volume on hope, Pieper offers a beautiful exposition of the nature of the biblical understanding of "the fear of the Lord," showing how it is precisely the attempt to eliminate *this fear* that undermines the pursuit of wisdom and "because presumption shuts out fear, it also shuts out the virtue of hope."[17]

And this theological virtue of hope is of course intimately tied to the theological virtues of faith and love. We cannot see the world as it really is apart from faith, hope, and love. And this returns us to the fundamental question of how things are, the reality that constrains what we can really see. Pieper acknowledges that the *confession of hope* is often made amidst suffering and in full recognition of the pain of the world.

> Such certainties all mean, at bottom, one and the same thing: the world is plumb and sound; that everything comes to its appointed goal; that in spite of all appearances, underlying all things is—peace, salvation, Gloria; that nothing and no one is lost; that "God holds in his hand the beginning, middle, and the end of all that is."[18]

Many of us will have difficulty making this confession with Pieper. Some of us have found ourselves living at the mourner's bench. Some of us have come face to face with the horror and sorrow of the world. Some of us have come to this place by walking on a very thin edge that does indeed falter between presumption and despair. Pieper has a word for us.

> How can we praise contemplation of this earthly creation when the ages, the present age and probably all ages, have been full of disorder, frightful injustice, hunger, painful deaths, oppression, and every form of human misery? . . . Is this anything but flight from the real world, an attempt to render horrors innocuous, a form of self-deception and unrealistic idyllicism? Ought not a generous person who does not care to deceive himself about what is going on in the world day after day—ought not such a person have the courage to renounce the "escape" of happiness [borne of hope]? . . . [N]o one who thinks of the world as at bottom unredeemable can accept the idea that contemplation is the supreme happiness of man. Neither happiness nor contemplation [nor the theological virtue of hope—*my inference*] is possible except on the basis of consent to the world as a whole. This consent has

little to do with "optimism." It is a consent that may be granted amid tears and the extremes of horror.[19]

And yet, it must be remembered that this *is* a confession, and a confession of faith, brought about by love, grounded in humility, issuing forth in hope, illustrating once again the unity of the virtues.

Humility and hope begin in the recognition of the realities of both the struggles of our world and the grace of the One who has not abandoned us to ourselves, either in presumption or in despair. In such a place, let us hope that we *might* be able to begin to hear and to speak the words of St. Paul to the Romans anew:

> And not only this, but we exult in our tribulations, knowing that tribulation brings about perseverance, and perseverance, proven character, and proven character, hope; and hope does not disappoint because the love of God has been poured within our hearts through the Holy Spirit who was given to us. (Rom 5:3-5)

IV

And this is why I find being a Christian, a farmer, and an intellectual so mutually edifying. The farm presents to me over and over again the joys and constraints of a reality that is really there—despite my best intellectual attempts to pretend that it (or life in general) is other than it is. Despite the best efforts of my time, money, and technology, the heifer will not calf, and the ewe's long-awaited lamb dies. Weeds emerge where there should have been fruit. The pond dries up in the drought, and the buzzards congregate on the dock to collect the fish that have died as the waters evaporate. The family dog kills the chicken, aphids ruin the squash, and raccoons take all the sweet corn.

And yet.

The spring brings baby chicks and goslings. The grasses return to the pasture. Cherokee Purples, Arkansas Travelers, and Early Girls, accompanied by jalapeños, serranos, and "long greens," all make their way back through the Texas heat. The miracle of birth occurs again and again—calves, lambs, piglets, and kittens.

They will not all make it. There will be more tragedies. But there will be blessings as well. And this is the case both on the farm and in the city. We all live in the tension between presumption and despair.

This is the complexity of the world and the challenge of neither taking ourselves too seriously nor abdicating our responsibility as stewards of a world we did not create and only possess as gift. And this is why I find being a Christian, a farmer, and an intellectual so mutually edifying—and correcting.

NEW GUINEAS

For the last few years, we've bought guinea fowl in the spring to help cut down on the grasshoppers, mosquitoes, ticks, and slugs. And they are supposed to be great for chasing off snakes. (Either this part isn't working, or if it is, we'd really be in a heap of trouble without them.) But I've had a hard time finding them this year, and when we have found them, they've been pretty expensive (five or six dollars each for week-old keets). Today Andrea sent me an email of some that were for sale (three dollars each) in a neighboring county. I called the guy this afternoon, and after dinner we drove down there to get a couple dozen.

This guy's farm was out in the country. And I don't mean like how we live in the country. I mean in the COUNTRY. After the pavement and the blacktop ran out. Beyond the low-water crossing. Giving directions, he said, "After you take a right at dat light and go a few miles, you gonna come over a big rise, and then come WAAAY down into dees bottoms. You can't miss it." When we finally found it, it was fantastic. He and his wife were sitting in lawn chairs under the shade of a large barn, surrounded by a menagerie. They had virtually everything. Cattle, sheep, goats, chickens, peacocks, doves, pigeons. They were also breeding Anatolian Shepherd guard dogs (they had eight new puppies). They had more pens and barns and chicken coops than I would have imagined. No two pieces of lumber or wire mesh matched, but it was all clean as could be. They showed us around with pride and explained what they were doing with all of the different animals.

Concerning the pigeons, he said, "People buy pigeons and don't hab da sense to keep 'em penned up. What do dey dink dey is goin' to do? Fly around da house and sleep on yo porch? Dis old boy from

San Antonio come up here and bought thirty-five pigeons for ten dollars each. And I told him you gots to keep 'em penned until they has der own chicks. Did he do it? Nope. Two weeks later ever last one of 'em was back here at MY house. Well, I caught 'em all for him and called him up and he come back up here and get them, and two weeks later dey was all back here safe and sound. I called him back, and he said, 'If those damned birds want to live at your house so bad, you can just have 'em!'"

He was asking us where we lived in Crawford (they live about twenty miles away), and when we said, "Compton Road," he said, "You mean Jack Compton Road?" (Jack was the farmer who built our house years ago.) Turns out he had known Jack and knew right where our house is. He said, "Jack was always giving me shit. One time I was plowing dis ole boy's field and der was dis big ole log. I got off the tractor, and when I tried to kick it away, a big ole rat ran out from it and then ran up my pants leg. I jerked off my britches and was just a shakin' 'em in the air, and here come ole Jack. By da time I got to da feedstore, everbody in Crawford knew about me standing beside da road waving my britches around over my head."

Inside the chicken house he had a bunch more pens and a brooder and an incubator, and we went in there and got two dozen keets. ("And I thowed two or three extras in there in case some of dem don't make it.") We boxed them up and brought them home. Now they're in our brooder, and we're off to the races with guineas again.

SKUNKS

There are many predators on a farm: coyotes, foxes, hawks, opossum, raccoons, and others. Some of these are more dangerous, but none strike fear in the heart as much as do the skunks. Skunks can be extraordinarily troublesome on the farm. They will make a mess of the garden, they love to eat eggs, and they will even attack chickens. But it is of course the spray and the smell they leave behind that make us quake in our boots.

We had experience with skunks even when we lived in suburbia. Our backyard ran down toward a ravine, and we would often get skunks and opossum in the backyard. At the time we had a Boxer named "Molly" who would invariably tangle with these vermin. Many times she would find one and start barking and barking. Our neighbors even called the police on her a couple of times. (And people wonder why we moved to the country?) Late one night after Andrea had gone to bed, I stepped outside and smelled the familiar odor of a skunk. Knowing Molly's proclivity for going after these creatures, I thought I would make a preemptive strike and bring her in the house *before* she got sprayed. When Molly came in on other occasions, we usually kept her in our bathroom that contained our closets. I had no sooner gotten her through the living room and into the bathroom before I hear Andrea call anxiously from the bedroom, "What's going on? Is something on fire? Scott, the house is on fire!" "No, no, everything's fine. I'm just bringing Molly in to keep her from getting sprayed by a skunk."

Andrea jumped out of bed. "She's already been sprayed! What's the matter with you! Are you crazy? The whole house is going to stink!!!" She was hopping mad and she was right. About all of it. Molly was only inside for a couple of minutes but it was long enough to infect the house. The entire downstairs now had skunk smell in it. We got the fans going, the candles burning, and the windows all open. We bathed Molly first in tomato sauce and then with a mixture of hydrogen peroxide, baking powder, and soap that Andrea found on the internet. Nothing worked.

The skunk smell was in the house—and in all our clothes—for a couple of weeks. We washed and rewashed the clothes. We tried everything. It was just horrible. Andrea read that some people have to throw away their clothes and get the house disinfected. Our children went to school and heard other children complaining, "Oohhh, what's that smell?" One of our sons came home and told us he sat at his desk thinking silently, "It's me." We've earned our fear of skunk stink.

Once or twice on the farm, we've had a skunk get under the house. The porch smell was atrocious. We could close up the access under the house, but we feared that we might inadvertently lock him in. A neighbor said that we should close up the access only after it got dark since they are nocturnal hunters. In the meantime we should scatter baby powder near the access and see if we get any tracks going out. We didn't find any tracks, but each time the porch smell slowly dissipated, so we believed we may have exiled him.

This spring we have had a number of skunks on the farm, and we had seen one or two go into the barn closest to the house. We thought that there might be a skunk in there, and we finally saw one burrowing his way under an old feed counter that we use in the barn to keep tools. And the skunk had made a home underneath it. Great. What do we do now? The counter is twelve feet long and four feet high. Moving it is very difficult under any circumstances and terrifying with a potential skunk den underneath it. We tried the "night-time block" approach, and he simply dug a new hole. Trapping him was no good. I was certainly not going to carry a cage with a live skunk in it. If we shot him in the barn, the barn would stink for weeks. It was clear that we would need to wait until we could shoot him outside.

One Saturday afternoon while collecting eggs, I saw the skunk coming toward me from the pond along the fence line. I tried to call Andrew on the cell but he wouldn't answer, and the skunk was quickly approaching. I called Andrea on the cell while yelling at the skunk to keep him away. I had a basket full of eggs in one hand and the cell phone in another, and I was trapped between a latched gate (that I had no free hand to open) and a rapidly closing skunk. Holding the phone between my head and shoulder, I picked up a stick and began to beat on the fence and yell at the skunk to get away while also telling Andrea, "Tell Andrew—get away!—to get his gun—hee-yah!—the skunk is outside." "What?!?!" "Get away! Get away!—I've got the skunk, and he's headed toward the barn!—Yaah!—Get Andrew's gun!"

Our eldest son, Benjamin, home for spring break from Notre Dame, appeared first with a shotgun. I had retreated through the gate but had not been able to stop the skunk's progress toward the barn where he had made his den. Screaming and waving my stick, I'm trying to stay in front of the skunk (for obvious reasons) to block his path to the barn but without getting too close to the skunk (again, for obvious reasons). But I don't want Benjamin to shoot him with a shotgun because it will blow a hole in the side of the barn and make everything stink. "No, no, not a shotgun!"

Benjamin was wearing a new sweater that his girlfriend gave him. Andrea appeared and told him, "If the skunk gets you, that sweater is going to be ruined." Benjamin immediately strips down to his boxers (apparently the pants were new too). Andrew, who had been showering, now appeared with wet hair and wearing only his shorts, armed with his .22. The skunk has hidden behind a trash can and some boxes at the edge of the barn. He's about ten feet away from his destination and safety. He is surrounded by the three hapless hunters, including two virtually naked young men with guns (one of whom is pasty white, having recently arrived from Indiana where the sun hasn't shone for months) and a middle-aged philosopher who is (mercifully) clothed but armed with only a stick and a basket of eggs. Andrea and the girls watch in horror from the mudroom window. We look like we're auditioning for a redneck reality TV show.

My goal is to drive him far enough outside of the barn that we can shoot him in the open. We'd been told that the only way to shoot a skunk and *not* have him spray is to hit him in the head. I'm shouting instructions to the boys, who are arguing about who has the best shot. I want Andrew to shoot him in the head with the .22, but in truth nobody has a good shot. So I'm trying to pull the boxes and trash can out without getting too close to him. But whenever I move an obstacle, he moves behind another. Finally Andrew gets his shot and takes it. Success.

But Pepé Le Pew got the last word. And he would not be forgotten for many days.

RATTLESNAKES

Wednesday was April 1. And we killed three rattlesnakes at the farm. No joke.

The first was Wednesday morning. I was getting dressed and looked out the bedroom window to discover a three-foot rattlesnake coming through the grass about five feet from our bedroom window. I got the shotgun and took care of it. But we were a little "rattled." In the almost four years that we've had the farm, we've seen very few snakes and no rattlers at all. A high school boy down the road had said that he killed one by the road in front of one of our pastures, but we'd never seen any.

I was getting ready to go, and we were taking the dogs out through the back door. Andrea and I had both walked through the door, and I'd come back out when I heard THAT sound. The rattle. I turned and saw another one wrapped around the electrical box at about the height of my head. We had both walked right by it, not two feet from our heads. Now what do we do about this one? We can't shoot it. I'm afraid that if I knock it down, it will simply hide under the deck. So Andrea and I stand outside and wait. We've traded the shotgun for the .22 (in hopes that it will make it to the ground where I can shoot it without shooting up our house). Finally it moves its head behind one of the steel cylinders that contains electrical wires. Now with his head on one side and his body on the other, I take a hoe and trap its head against the steel pipe. It's flapping around, the rattle is going crazy, and I'm hanging on for dear life. My angle is bad and I can't cut off the head with the hoe, but I might be able to crush its skull, provided I don't pull the pipe out of the box. Andrea goes and gets Andrew's hunting knife. When she comes back, I give her the handle to the hoe to hold the head tight and I cut into it.

When it finally seems to give in, I let up a little and fling him into the yard where I can use the hoe to take the head off properly. Another three-footer.

All of this takes about an hour, and I've never killed a snake before while wearing a tie. We're both a little shaken up, but we are comforted by the memory from Huck Finn that rattlesnakes always come in twos. That's bound to be it. We pledge to be more careful and to stop going to the barns in sandals, and I go to school.

After Andrea picks up Andrew from school, he's keen to cut the rattles off of the two snakes. While he's in the side yard cutting off the rattles, Andrea is on the deck. She looks up and there is another one. This one is wrapped around the floodlights. By the time she can get me on the phone and get Andrew to the back yard, number three had made his way to the deck, coiled by the back door. Andrew has the .22 and wants to shoot it. I'm telling Andrea, "Don't let him shoot the house. If he misses he'll shatter the glass of the French doors." "Tell Dad I won't miss. I can hit him!" "No! Don't shoot him until you can either get it in the yard or at the base of house by the concrete skirt." Now Andrea pipes in, "The snake is moving. It'll get under the deck, and we'll live in fear till it's out—Andrew's got to shoot it." Next I hear three quick shots followed by general chaos. Did they get it? What's going on? A couple more shots. "I got him! I got him!" "Is he dead?" "I can't tell; he's still moving." "Get him in the yard and use the hoe." "I think he's still alive. Where's the shotgun?" "You don't need the shotgun, use the hoe!" "I got him, I got him!"

Andrew did get the snake, and he also put a nice size hole in the metal threshold of the door.

In the last two days, we haven't seen anymore, but the story of our snakes has made the rounds in Crawford. The common conclusion is that there was (is?) a den nearby, and these three came out to sun. Neighbors have stopped by, and everybody has a theory. "The biggest ones always come out first. Look for the babies next." "Once they come out, they scatter. They'll all be gone in a week's time." Nobody knows why they were on the side of the house. Rattlers are definitely NOT climbers. Were they going up for a warm spot?

Chasing food? Maybe there were coming DOWN. If so, where were they? Are there anymore? We're wearing boots and packing heat.

The best (or worst) theory was articulated by Doug. Andrea and I had already discussed this option but tried to put it out of our minds. Doug said, "You know, Moore, now that you've rerouted your septic system, that old shithole under your garage is now empty. That's probably a perfect place for a bunch of snakes to hide out for the winter." Great.

Before this is all said and done, we may end up jackhammering through our garage after all. Stay tuned.

DEAD LAMBS

Gene Logsdon said in one of his books, "If you raise sheep, you will have dead lambs."[1] I didn't want to believe that, but it's true.

We have a small flock of Tunis sheep. Tunis are a heritage, dual-purpose breed of sheep. They are called "redheads" colloquially because when they are born, they are a deep red color. When their wool comes in, it is a beautiful color of ivory, but even then their heads and legs remain the dark red color of their birth.

The first lamb we lost was accidentally killed by an aging, incompetent veterinarian. We had learned how to dock the tails of young lambs with our first set of twins. Within the first two or three days after birth, you simply put a very small rubber band an inch or two from the base of the tail. An elastrator is used to stretch out the rubber band and insert it over the tail. Within a couple of weeks, the portion of the tail beneath the band will simply fall off. This method is used both for docking the tails and for castration. But it must be done pretty quickly after birth.

Our second set of twins was much weaker than the first. Their mother had never given birth before, and she did not nurse her lambs as she should have. Andrea had to bottle-feed both lambs, and we weren't sure if they would even make it. By about day seven or eight, they were getting stronger, and if I held the mama, Andrea could hold the lamb and make sure that it was able to nurse some. But we had not docked their tails because we didn't want to add any additional stress to these weak young lambs.

This was very early in our farm adventure, and we did not yet have a large animal veterinarian that we knew well and trusted. Since our lambs were now over a week old, we thought it best to take them to a vet to get their tails docked. We called around town, found someone who saw sheep, and took our two little lambs in. The doctor said it would take a couple of hours, and we could pick them up that afternoon.

When Andrea returned to collect them, the elderly doctor said nonchalantly, "One of them didn't make it." What? How is this possible? "Sometimes they just don't make it. But I won't charge you for that one." We never knew what really happened. The one that died ("Penny" we had named her) was the larger and stronger of the two. Art Hunter, our ag teacher at the high school and our resident expert on all things agricultural, speculated that he either gave her too much anesthesia or perhaps he cut the tail off too close to the body, and she bled to death. Or maybe she just died. It happens.

For several days I contemplated all kinds of ridiculous actions. We could sue. We could go back down there and demand an explanation or compensation. We never even saw the body. Maybe she's alive, and he just stole her! At the very least I wanted to file a report with the Better Business Bureau. In the end we did nothing. But we were devastated. This was the first animal death we had on the farm. And we knew that we had taken her to this man we knew nothing about, and we were ultimately responsible.

"Francis" was our second lamb to die. Andrew first discovered him close by his mother's side in the back pasture the same weekend that Pope Francis was selected to be pope. Our son Benjamin was studying in Rome that year and had kept us posted on all the goings-on, and we had all been excited about the pope's election. It only seemed right to name our new lamb Francis. At first he was strong, and he seemed to be nursing well. But in a few short days, we saw that something wasn't right. He would sleep in the corner of the stall, and even though his mother would try to nudge him to get him up and to nurse, he just lay on the hay. By day ten he had died.

We had a similar experience with a little unnamed ram who was born in February of 2016. Over the course of four days (while I was in Athens), five lambs were born to our ewes. Andrea was attending to them and sending me texts and pictures, and it was magnificent. The last of these arrived the very night I returned, and once again Andrea and I had to help with the delivery. But one of the little rams didn't last a week. We found another one in the corral that we didn't even know was coming. It had died before we could even discover it. Another one was taken by a coyote. We found a mangled carcass behind the corral near the pond.

We've had about twenty lambs born to our little flock on the farm, and five of them didn't make it. As with most things, Gene was right. "If you raise sheep, you will have dead lambs."

ALEXANDER McCALL SMITH

Alexander McCall Smith is one of my favorite contemporary authors. A former professor of law and medical ethics, he has authored dozens of books, from serious academic works on the law and legal philosophy to hilarious serials poking fun at the pretentious bourgeoisie. He is best known for his series *The No. 1 Ladies' Detective Agency*, in which he details the adventures of Botswana's most delightful detective, Mma Precious Ramotswe, and her family, friends, and staff. In addition to the Botswana books, he is currently maintaining two series set in his beloved Edinburgh: *The Sunday Philosophy Club*, in which we meet Isabel Dalhousie, a philosopher who solves mysteries, and *44 Scotland Street*, a serial featuring the lives of several families and individuals who (originally) shared a collection of flats at the Scotland Street address. There are other series as well and a number of stand-alone novels and short story collections.

At first glance it might seem that McCall Smith's books are unlikely instances of agrarian literature. The vast majority of his stories are set in urban areas, especially the cities of Edinburgh and London. Even the *No. 1 Ladies' Detective Agency* series is set in the booming city of Gaborone, Botswana. But on closer examination, it becomes clear that there are numerous agrarian themes that repeatedly appear in his books and stories.

The most obvious of these is the profound love and appreciation for cattle that his African characters have. In every Botswana novel, we are reminded that cattle are the traditional source of wealth and security in that country. Precious Ramotswe was able to start her detective agency because her beloved father, Obed Ramotswe, not only left her a large and healthy herd, but he also taught her how to recognize and appreciate good cattle. Some of my favorite moments in the Botswana stories occur when Mma Ramotswe drives her tiny white van into the vast Botswanan countryside and reflects on the beauty of the cattle she finds there. There are too many examples of this to mention, but I have often found myself in a similar position.

As you drive across Texas country roads, cattle are ubiquitous. And as such it's so easy to miss them. It goes without saying that our own adventures into the cattle business have changed the way we view cattle, but it's also the case that on occasion I have done exactly as Mma Ramotswe. I have pulled over to the side of the road just to watch the cattle. Perhaps it's a cow with her calf. Or perhaps some slightly older calves are jumping and playing with one another. It might merely be a great bull relaxing in the shade of a hackberry tree or an exuberant young bull pacing one fence line with an eye to joining the heifers on the other side. In the chapter "In Defense of Watching Grass Grow" I quote Wendell Berry's Andy Catlett, who confesses that it took him a long time to appreciate his father's favorite sight: cows grazing on good grass.

We have begun to learn the common breeds that are found in our part of the world, and it's great fun to identify the differences between them, especially some of the more common crossbreeds. "Are those Brangus, Andrew?" I might ask as we pull to the side of the road. The teenager's response is definitive: "Oh no, Dad. Come on. Those are Bradfords. Anybody can tell that."

But for McCall Smith, it's not just the cattle. He loves animals and those who love them. His two serial collections, *44 Scotland Street* and *Corduroy Mansions*, prominently feature the lives and adventures of dogs. Angus Lordie's dog, Cyril, has a gold tooth, drinks beer from a saucer in the Cumberland Bar, and has had numerous adventures of his own. And Freddie de la Hay is virtually the star of the show in *Corduroy Mansions*. Indeed, the second volume is entitled *The Dog Who Came in from the Cold* and chronicles (among other things) Freddie's service to the British nation as an agent of MI6. But there are of course many other dogs in McCall Smith's work. There is Mealies, the snake-catching dog, in *The Minor Adjustment Beauty Salon*; Zebra, whom Fanwell found and could not give up; Lubka, the Labrador Retriever in *A Distant View of Everything*, whose eyes match that of his owner, Peter Stevenson; and of course Peter Woodhouse, Dog First Class, in the novel of the same name.

And we must not forget Brother Fox, the beautiful red fox who regularly appears beneath the rhododendron bushes in Isabel Dalhousie's garden. Against the wishes of her neighbors, Isabel feeds him from time to time, and she even calls in a veterinarian to tend to a nasty wound he has in *The Lost Art of Gratitude*. Brother Fox is a reminder to Isabel of the communion she shares with nature—even in the midst of the city.

McCall Smith also loves the countryside. Whether it's Isabel and Jamie venturing through the Highlands to the Hebrides, Bertie and Stuart getting lost in the Pentland Hills, or Angus Lordie and Domenica romping through Tuscany, over and over again McCall Smith treats his readers to the glory of the trees and the fields and the flowers of the land. Even the austere beauty of the Botswana bush country takes on a significance in his presentation of it.

McCall Smith has a deep appreciation for farms and farming, which comes through in many of his novels. Mma Potokwani's Orphan Farm is no Dickensian orphanage. It is a working farm (despite the fact that much of the equipment barely works) in which children live in bungalows and are lovingly cared for by house mothers. Big Lou not only regularly recalls the lessons she learned growing up on the farm in Arbroath, but she also maintains a special interest in agricultural matters, like the rare breeds of pigs raised at her friend Alex's farm, "Mains of Mochle." In both *La's Orchestra Saves the World* and *The Good Pilot Peter Woodhouse*, English farms become the setting for renewed hope and friendship during the difficult years of the second World War.

And in *The Forgotten Affairs of Youth*, Isabel makes an explicit affirmation of the centrality of farming while driving through the Scottish countryside.

> She drove past a large field of hay that had been cut a day or two previously, the circular bales dotted about where the harvester had disgorged them. She saw a tractor halfway up a slope, driven by a man who was waving to another man on the ground; she passed a field full of pigs with their curious, domed pig arks like the tepees of some tribe

of plainsmen. She thought: All *this* happens to support all *that*—that being the life of the cities, all those people who were ignorant or indifferent to the life of the countryside and to their agricultural roots. Music and art and philosophy are ultimately based on the premise that this man on his tractor, and these pigs, and the swarms of bees that fertilise the crops, will all continue to do what they do. And every philosopher, no matter how brilliant his or her insights, needs a portion of this field—how much? Half an acre?—to support him if he is to survive.[1]

In *The Minor Adjustment Beauty Salon*, Mma Ramotswe visits a farm outside of Gaborone on one of her investigations. The narrator tells us, "She had not been to this farm before, but the scene seemed both familiar and peaceful. This, she thought, is how you should live, if you possibly could: with your cattle around you, with the land beneath you, with this air about you."[2]

I think Mma Ramotswe is correct, but of course she usually is.

HOW MANY CHICKENS? HOW MANY EGGS?

Finding the right number of hens to produce the desired number of eggs is harder than it may appear. In theory one ought to be able to know approximately how many eggs a hen of a particular breed should lay in a week and then make one's calculations accordingly. But it almost never works out that way for us. We invariably have too many eggs, too few, or all too often too many eggs that we can't use and not enough that we can.

"They" (Das Man) say that you should be able to count on a laying hen giving you *at least* eighteen months of productivity. And since it takes on average about twenty-five weeks for a chick to develop and start laying, that means that you need to able to replace your flock with mature birds every two years.

Of course, there are many, many variables. For starters, many hens will produce for much longer than eighteen months. We've had a few hens that kept laying for three years, though their production was obviously diminished. And over the course of the year, you always lose a fair number of hens. Some get taken by unknown predators (coyotes, raccoons, chicken hawks), and some die at the hands (jaws?) of the family dog. Others get sick and die for no obvious reason at all.

Even if you know how many are going to live (which you don't), there are many different ways the "average egg calculation" can go wrong. For starters, they lay at different rates depending on the time of year, temperature, amount of light in the day, and most importantly, how stressed they are. If they are too stressed, they can stop production altogether. And there are many things that can stress a hen. In our experience the worst thing that can happen to a flock is for a predator to get in the coop. I've already told the story of how our neighbors' dogs got into our chicken yard and killed about half the flock. All of the birds that lived were so traumatized that they stopped laying for a month. During this time we had to buy eggs at the store. I was humiliated. From then on we kept two separate coops.

But most chicken stress is not nearly so dramatic. Mostly they get stressed by other hens. (Imagine that.) If you introduce new birds into your flock, they will need to establish a new "pecking order." This causes considerable stress. If you change their feed or their environment or clip their wings to keep them out of the garden, stress. (In truth, it's not the clipping of the wings that stresses them but rather the being chased and caught by humans. One chicken running around a coop squawking will get the whole flock upset.) Roosters (no surprise here) can be the source of a great deal of stress to the hens—and not just when they are bothering the girls. And of course if the coop is too crowded or they don't have enough food or fresh water, they get stressed.

Sometimes figuring out why they are not laying many eggs becomes a detective problem. One summer during a particularly bad drought, egg production in both coops almost completely stopped. Some days we got absolutely nothing—and this from about fifty laying hens who had been giving us two to three dozen eggs a day for weeks before. Why? Well, it had been unusually hot, even for Texas, with many days over 105 degrees. (That summer we set an all-time Waco record of 114 one day.) The drought had also driven an increase in chicken snakes into our coops. Every time production dropped significantly, we found a new snake. One day we even had two big chicken snakes in one coop taking our eggs! But at the height of our egg drought, we stopped seeing snakes. There were no cracked eggs, so it probably wasn't a raccoon or opossum. And since both coops had stopped producing, we figured it must be the heat. But then on top of a large round bale inside the pig pen, we found a nest with eight eggs. So it wasn't only the heat. Why had they quit laying? We may never know. After a couple of weeks, production slowly resumed.

This makes it sound like keeping chickens is a constant battle with poultry stress and predators, but that is definitely not the case. In most cases you secure the coop and establish a pattern over time, and some days you get lots of eggs and some days not so many.

For the first couple of years, we simply gave away all of our extra eggs. We didn't really know what we were doing. We had extra eggs, friends and family gave us their old egg cartons, and we just filled them up and passed them out. I might take five or six dozen to the office, or Andrea would take several dozen to a luncheon with friends. My parents would pass them out to their friends, and we would take a bag full every so often to church. People often offered to pay, but we usually said no. We enjoyed giving out eggs.

Growing up and traveling overseas with my professor parents, as a child I had always been amazed at how the eggs for sale in European grocery stores weren't refrigerated. Didn't they go bad if you left them out? Did the English know that they were buying rotten eggs? There was much about England that was inexplicable to me. Could it all go back to rotten eggs? How could they just put eggs out on the shelf? It wasn't until we started raising our own chickens that we learned that a freshly laid egg has a natural protective coating on it that will keep it fresh for a couple of weeks. The egg only has to be refrigerated if you CLEAN the egg and scrub off the protective covering. Once it has been refrigerated, it has to stay refrigerated.

So now you have a dilemma: you can have clean eggs that have to be refrigerated or (potentially) dirty eggs that can be left at room temperature. We have usually opted for the latter with the proviso that we try to collect eggs two or three times a day so that they are not *too* dirty.

One Christmas Andrea had a "Benedict Farms Fresh Eggs" stamp made for us, and from that time on, we started ordering new cartons and putting our stamp on it. However, if you want to sell your eggs at a farmer's market, then they have to be clean and refrigerated. Our county health department even specifies at what temperature the eggs have to be kept, which means that you cannot sell eggs from a mere cooler in the truck. You have to sell them from a cooler in which you can regulate the precise temperature (otherwise known as a refrigerator). This means you need to be able to plug your, er, cooler into an electrical outlet. And this of course means that the farmer's market has to HAVE available electrical outlets.

There are two main farmers' markets in Waco, one with outlets and one without. At our beautiful, vibrant Downtown Farmers Market, they offer booths with outlets. But we don't really have enough eggs to justify paying their vendor fees. The older, smaller farmer's market has much more modest fees, but the health department won't let their vendors sell eggs because they don't have electrical outlets.

Our solution was simply to take several dozen eggs to our church office and let people take what they need and leave a few dollars in an envelope. This "egg money" goes to help with paying for chicken feed and other small matters. After we had been doing this for awhile, we were thrilled to discover that this is one of William Cavanaugh's proposed practices to help communities avoid *Being Consumed* by the larger consumer society. Cavanaugh's book is superb, and it is filled with examples of concrete practices that can help us work together to meet community needs.

But this doesn't solve the problem of "How Many Eggs?" We remain a rather unreliable source of eggs. Many weeks we run out, and by the time we get more to the church office, some of our friends have had to buy elsewhere. And occasionally someone will get a "bad egg" from us, and that definitely puts a damper on their enthusiasm for the whole farm-to-table experience. I recently read a website that said, "If you're going to go into business, you've got to fix these problems. People have to know that they can count on you having eggs for them and that they will always be good." But that's precisely what we don't know.

OCKHAM, IRIS, AND THE SHOW CATTLE

We survived the Heart O' Texas Fair & Rodeo. On the surface we did more than survive; we almost flourished. The two heifers we took to the fair both got second place in their respective classes, and we have good reason to believe that one of them has some real potential. Moreover, neither of the cows got loose and ran around the parking lot, which is not to be taken for granted. But the claim that we "almost flourished" needs to be taken with several pounds of salt. It's true but deeply misleading, and it is typical of the consoling fantasies that we tell ourselves in the midst of life's little tragicomedies. Iris Murdoch believed that "any story which we tell about ourselves consoles us since it imposes pattern upon something which might otherwise seem intolerably chancy and incomplete." And later she writes, "The difficulty is to keep the attention fixed upon the real situation and to prevent it from returning surreptitiously to the self with consolations of self-pity, resentment, fantasy and despair.... It is a *task* to come to see the world as it is."[1]

"It is a task to come to see the world as it is." In truth both cows had serious difficulties. The young heifer ("Boogie") had been the focus of most of our attention in the weeks leading up the fair. She is the first heifer that we have completely and satisfactorily broken (so we thought). In the last two weeks before the show, she was walking very well for us. Andrew could lead her in large loops through the pasture, and she would even set up well in a show stance. But at the fair, she was a rambunctious mess. She fought Andrew all around the ring. Andrew never lost his hold on her, but it was pretty bad. Andrea was sitting in the stands and said she heard one person sitting behind her say, "That heifer ain't broke" and another say, "Look at that boy sweat—he's working hard." There were only two heifers in this scramble class, so Boogie did get a red second-place ribbon. We got her back to her stall and decided that we would NOT show her the next morning in the second competition.

The older heifer ("Honey") also got second in her class of nine, so we were genuinely pleased with this result. She was also frisky, though it wasn't nearly as bad as it had been with Boogie. But it was far from ideal. She tugged and pulled and wouldn't stand still. When the judge commented on the cattle, after praising but making several critical comments about the one he placed first, he said of ours, "The only thing this one I've placed second needs is to learn to like being a show heifer." What I don't think the judge saw was that when we were in the holding pen about to go in to the ring, Andrew was swapping out the rope halter for a leather one, and she got away from him. For several minutes she was loose in that small pen, running around, literally kicking up her heels, and it took several minutes for us to corral her and get her halter back on her. It was fairly humiliating. But at least she looked good when we got her in the ring.

It was at these same Heart O' Texas Fairgrounds in the spring when "Silky Smooth" slipped and fell while coming off of the trailer, and thus got spooked, bolted, and spent more than an hour running around the parking lot. This was truly one of the worst hours of my life as I chased her from one end of the parking lot to the other with visions of liability and destruction racing through my mind. Our $3,000 cow had almost made it to the busy four-lane Bosque Boulevard before being cut off by a speeding truck that herded her back toward the fences near the arena. At one point she ran through a group of Waco High School Band members who were on their way to practice. I heard one old rancher say, "Them band kids scattered like a flock of seagulls." Ultimately she was lassoed by a couple of cowboys from the back of a flatbed truck and trapped in a stranger's trailer. When we tried to move her to our trailer, she was mad as hell, snorting and kicking and charging the side of the trailer. With twenty people looking on, I moved our trailer behind the stranger's and tried to coax her into ours. When we finally got her moved, she still had the cowboy's lasso around her neck. I began to try to get the rope for him, but he said, "Mister, you can have it. I wouldn't go in there for love nor money."

So a couple of rambunctious cattle that don't get loose and still win ribbons doesn't sound so bad.

But all of this made me reflect on our brief foray into the cattle business. We have owned six head of cattle in the last three years: five Beefmaster show heifers and one Jersey calf that Andrea bottle-fed from the time she was two weeks old. Of the five show heifers, none of them have really been successful. The first, and best, was "Bessie." Beautiful and docile, she was almost entirely broken by the time we got her from the ranch. We had to work with her to get her to walk, but it never crossed our minds that she might take off and run down a busy city street. In every competition we entered, Bessie placed. She was never a champion, but she was always in the top group and always calm and peaceful.

We tried for almost two years to breed her. We left her with good bulls, and we tried artificial insemination, but nothing worked. Finally the vet did an ultrasound on her, and she simply didn't have all the parts she needed. The job of a cow is to produce calves, but Bessie wasn't up to it.

The second heifer we bought was "Scarlet." We (or rather I) bought her at an auction at the Houston Rodeo, and she had great bloodlines. When the rancher from whom we bought her brought her to the farm, he sheepishly admitted that he didn't think she would ever be a great show heifer, but she was good for breeding. Her sire and her dam were both champions, and while we shouldn't get our hopes up for the show ring, she would produce great calves. Andrea, who had not been consulted on this latest great idea, cut her eyes toward me and nodded knowingly. Two years later and still no calves from Scarlet—though I might add, she does have all the parts. At the moment Scarlet is being bred with straws of prestigious bull semen we ordered from a catalog and paid to ship to our vet in expensive nitrogen-filled holding tanks.

And then of course there are Silky and Honey and Boogie.

I had wanted to believe that maybe we had just had some particularly bad luck. We've bought good cattle from reputable ranches, but we don't seem to be able to train or breed the cattle the way they need to be. And my conviction that we were just unlucky seemed more and more like one of Iris's consoling fantasies.

William of Ockham knew that the simplest explanation was always the most likely. And the simplest explanation was that we just weren't very good at raising cattle. Though we were constantly seeking advice, we just didn't know enough. We talked with cattlemen, read books and articles, and even ordered videos off of the internet, but we were still greenhorns. Moreover, we probably didn't spend as much time as we should have with them, and we certainly didn't have the facilities or expertise to do this well. Andrea put it most succinctly: "We just suck at this."

Was that it? Ockham's Principle of Parsimony reduced to the simple fact that we suck at raising cattle? There was bound to be more to it than this.

As I reflected on this problem, it seemed to me that there were at least four possible alternatives. Of course, the first and most likely (call it "A") was that we suck, or to put it another way, we lack the requisite *techne* to succeed at this endeavor. At the other end of the spectrum, the most consoling (and presumably least likely) explanation ("D") was that we were the victims of really atrocious luck. Things just hadn't worked out for us.

In between these extremes there were two other possibilities. First ("B"), there was the possibility that while it might be true that we suck at this, the larger context was that really only big ranches with paid, experienced cattle hands, and generations of leather-handed cattlemen could actually pull this off. And of course there is lots of evidence for this view. At every competition it's the big ranches that win. Even when the competitions are limited to teenagers, it's the grandsons and granddaughters of the successful ranchers who take all the awards. And these people not only know what they're doing, they have people who spend all their time perfecting this process. And frequently it's the hands who show up to make the beds, unload the cattle, feed them, and get them ready for the show, and then the granddaughter arrives and takes the cow into the ring.

The final ("C") explanation was that it wasn't the case that only big ranches could win but that only people who really know what they're doing and *spend most of their time doing it* are likely to be successful at this endeavor. We fall quite short of the mark on both counts. We have learned a lot, but we're still grossly inept at most of it. To see Andrea, Andrew, and me out trying to train the cattle to walk or trying to catch one that doesn't have a halter would make for a great reality TV show. By the time Andrew finally gets home from football practice, dusk is quickly approaching, and we are walking through the mud and the muck, intermittently coaxing and cursing the cattle to behave ("Come on, sweet girl, you can do it," and "Dang it, Honey, stop it!"), all the while encouraging one another with, shall we say, constructive criticism and helpful tidbits of advice gleaned from serious cattlemen ("Hold her head up"; "Don't give her so much line"; "Stay back, you're making her crazy"; "Remember what that guy said . . ."; "I know what I'm doing"; "Twist her tail"; "Don't get too close!"). Perhaps this is just a demonstrative presentation of what it means "to suck at this."

But we're not ready to give up—not yet, at least. We're holding out hope that we can breed these cattle, and perhaps—perhaps—if we get started with the calves early enough, we can do a better job of breaking them and then of showing them.

Of course, given that bad luck of ours—we might just have all bulls.

UPDATE: A year later we begin to see some success. Honey gave birth to a beautiful little heifer, and Boogie and "Annabelle" produced magnificent little bulls. See chapter entitled "Calves."

WENDELL, GENE, AND JOEL

ON THE DIFFICULTIES OF THEOLOGY AND AGRICULTURE

The story of how certain strands of Christianity in the United States became alienated from ecology and environmentalism is a fascinating one. In the nineteenth and early twentieth centuries (and long before there was a secular "environmental movement"), environmental stewardship and agrarianism were prominent features within both Protestantism and Catholicism. The Lord's Acre Plan, Rural Life, and Soil Stewardship Sundays became concrete ways for Protestant Christians to preserve God's creation and sustain both church and community. Among Catholics, the National Catholic Rural Life Conference began work in the 1920s, drawing on the strong traditions of Catholic agrarianism.[1]

Today, however, there are often tensions between "environmentalism" and certain strands of religious belief. The tensions and difficulties found in the theological reflection on agriculture can be seen in three of the most prominent voices in the contemporary literature of sustainable agriculture and cottage farming: Wendell Berry, Gene Logsdon, and Joel Salatin. Whether in fiction or in nonfiction, lyrical poetry or practical manuals, stinging cultural criticism or self-deprecating autobiography, each of these men have made a profound impact on the revival of interest in sustainable agriculture and cottage farming. Before Logsdon passed away in 2016, they were also friends and fellow laborers. They have praised and on occasion introduced the published work of the others. And while each of them also grew up in a religious tradition, they have taken very different approaches to the role theological reflection and religious practice might play in the work of farming. Taken together, these three farmer-authors exemplify the contemporary struggle to make sense of how theology and agriculture fit together.

Berry, a self-described "bad-weather churchgoer"[2] who goes to church when it is too wet to walk among the hills and the creeks, writes from within the Christian tradition, and yet he has been sharply critical of the ways in which the Christian Church has poorly equipped its parishioners with the skills or the insights to be faithful to the "gift of good land." Logsdon, the "Contrary Farmer" most noted for his voluminous insights and witty commentary on all of the practical matters of the farm, was a Catholic seminarian who ultimately left the Church. Salatin, though equally critical of the ways those in the pulpits and the pews have responded to the environmental challenges of industrial agriculture, describes and practices an understanding of farming and stewardship in the rich vernacular of the Christian tradition. For each of them, their differing connections (past and present) to the Church and the life of faith have played profound roles in shaping their understanding of the practice and vocation of farming.

My own reflection on faith and farming has led me time and again to the work of each of these self-described "crazy" farmers. (Salatin calls himself a "lunatic farmer," Logsdon published books and essays under the label "the Contrary Farmer," and Berry was known for his "Mad Farmer" poems.) For our family they have been some of our most cherished guides as we have sought to understand the difficulties surrounding the integration of theological motivation and insight into the practice of sustainable agriculture.

THREE FARMERS, THREE THEOLOGIES

Wendell Berry is certainly the best known of the three. Born in 1934 in Henry County, Kentucky, the author of more than fifty books, Berry is a prominent figure in the world of American letters and cultural criticism today. In his own self-description, he is preeminently a farmer. And yet he is also a poet, a novelist, a conservationist, an environmental activist, a cultural critic, and much more. He has won numerous awards, including a Guggenheim Fellowship and the Richard C. Holbrooke Distinguished Achievement Award for advancing peace through literature, received the National Humanities Medal in 2011, and gave the Jefferson Lecture in 2012.

Berry's novels and short stories are principally set in and around the fictional town of Port William, Kentucky, modeled on Berry's own hometown, Port Royal. Spanning more than a hundred fictional years, Berry has chronicled the joys and struggles of "the membership" of Port William. Berry's nonfiction runs to more than twenty books of cultural criticism and reflection. He is perhaps best known for his critique of the industrialization of agriculture, the rise of agribusiness conglomerates, and the decline of the family farm. His 1977 book *The Unsettling of America: Culture and Agriculture* remains a foundational text of modern agrarianism.

Gene Logsdon (1931–2016) farmed a small thirty-six-acre farm in Wyandot County, Ohio. In the words of Wendell Berry, his farm was "the best, the most productive, and the most beautiful small farm I have ever seen."[3] The author of thirty books and hundreds of articles, Logsdon is best known for his many books on the practical matters of the farm. The most famous is *The Contrary Farmer*, his defense of the significance and the viability of sustainable agriculture or what Logsdon calls the "cottage farm." He authored books on soil conservation, pond construction and management, growing berries, organic orchards, cultivating the woodlot, and how to use and respond to wildlife in the gardens and fields. In *Small Scale Grain Raising*, he instructs his readers on how to grow a "pancake patch." In *Good Spirits: A New Look at Ol' Demon Alcohol*, he discusses homemade beer, the backyard winery, and "how to make untaxed spirits legally." My personal favorite Logsdon book is *Holy Shit: Managing Manure to Save Mankind*.

Logsdon had a strong background in literature and philosophy. In addition to his seminary education, he also did graduate work in American Studies and Folklore at Indiana University. His love of farming led him to give up the university just has he had given up the priesthood, but even his practical farming manuals sparkle with references to theology and philosophy—though often as the punch line of his wonderful wit. For instance, in the aforementioned book on, er, manure, he has an entire chapter on "fork work"—the different types of pitchforks and how to use them. He concludes with these words:

> Trying to describe fork work with words or even with pictures can't substitute for actually doing it. Go at it slow. It is always best to have a companion who likes to talk. There is something about slowly, patiently forking manure into a spreader that encourages conversation almost as much as a sip or two of good whiskey. Philosophers would make great manure pitchers.[4]

The *New York Times* has called Joel Salatin (born in 1957) "the High Priest of the Pasture."[5] He burst onto the national consciousness in Michael Pollan's 2006 bestseller *The Omnivore's Dilemma*, but Salatin had been engaged in what he calls "the ecstasy of being a lunatic farmer" long before that. Pollan's book traced the "natural history" of four different approaches to getting one's food to the table: large-scale industrial farming, "big (or industrial) organic," small organic, and the forager. Salatin and his five-hundred-acre "Polyface Farms" in Virginia became the model for the small-scale "beyond organic" farm. As he was researching some of the criticisms that were arising from within Big Organic, Pollan kept hearing about a farmer in Virginia who was producing exceptional food. He called down hoping that he could get a chicken or steak FedExed to him in New York. Salatin refused. "I don't think you understand. I don't believe it's sustainable [. . .] to FedEx meat all around the country. [. . .] Just because we *can* ship organic lettuce from the Salinas Valley, or organic cut flowers from Peru, doesn't mean that we *should* do it, not if we're really serious about energy and seasonability and bioregionalism. I'm afraid if you want to try one of our chickens, you're going to have to drive down here to Swoope to pick it up."[6] Pollan not only drove down to get his chicken, he worked on Salatin's farm for a week to see how it was done. After his first day of work, Pollan noted, "thus far my principal conclusion was that in the event I survived the labors of the week, I would never again begrudge a farmer any price he cared to name for his produce: one dollar an egg seemed entirely reasonable; fifty dollars for a steak a steal."[7]

In addition to running a profitable and extraordinarily innovative farm, Salatin is the author of more than a dozen books and an increasingly popular speaker at Farm to Table conferences and Foodie festivals. His best-known work, *Folks, This Ain't Normal*, is a full-throated critique of both the Industrial and the Big Organic agricultural models, but most of his works offer suggestions for how small-scale farmers can actually make a living without giving in to chemical fertilizers or big ag centralization. These books have titles like *You Can Farm*, *Family Friendly Farming*, and *Your Successful Farm Business: Production, Profits, and Pleasure*. It should come as no surprise that Salatin and his unconventional farming techniques come in for both great praise and vicious criticism from the far corners of the agricultural world. Constantly at odds with both the FDA and the agricultural establishment, Salatin's mindset is succinctly captured in the title of one of his books: *Everything I Want to Do Is Illegal*.

At first glance one might believe that there is a mutual admiration society among these three contrarian farmers, as each has praised both the farming and the writing that the others have accomplished. Berry wrote a foreword or an introduction to several of Logsdon's books, including his parable *The Man Who Created Paradise*. Berry's praise of Logsdon is without parallel. In the foreword to the revised edition of *Living at Nature's Pace*, Berry describes Logsdon as "the best agricultural writer we have." And in the foreword to Logsdon's last book, the posthumously published *Letter to a Young Farmer: How to Live Richly without Wealth on the New Garden Farm*, Berry notes that the only diploma he has on his wall is a certificate naming him a member of The Contrary Farmers of America. "It is signed by Gene."[8]

In Berry's most recent book, *The Art of Loading Brush*, he breaks the boundaries between fiction and autobiography by telling his own story through the life of Andy Catlett, the character in the Port William fiction who most resembles Berry himself. (Both Andy and Wendell are born in 1934, both have lawyer fathers and brothers, both are writers, both leave the farm and the community, only to be "re-membered" back into it much later.) Berry describes how in about 1970 "Andy" met Gene Logsdon, and it was as if they had met each other long before. "They talked on for forty-six years, until Gene died, also in 2016, at the end of May."[9] He describes Gene as one of "the four allies he knew he had away from home."[10]

Logsdon also has a version of how the two met. At the time Logsdon was working as an editor at *Farm Journal*, a magazine in Philadelphia. A copy of Berry's slim volume of poetry, *Farming: A Handbook*, found its way to his in-box, covered up by press releases. Logsdon's incredulity turned to excitement as he read the poems. He stormed down to the managing editor's office to get permission to go to Kentucky to interview this farmer-poet. *Farm Journal* did not review poetry. "Corn-'n-Beans was god, generating the ad revenue that the magazine lived on. Editorial policy barely admitted that Kentucky's whiskey and tobacco existed. [...] If only this Wendell Berry lived in Iowa. Or Illinois. Then he could justify the story to the publisher." Finally the manager gave in. "Okay. But find another story to do while you're out there." Logsdon went, but he did not find another story. "Something happened between us that was more than friendship, if there can be anything more than friendship. We were cultural twins, farm boys who had grown up with similar experiences." They both had liberal arts educations, they both had known the trials of boarding schools, and they both had gone to the city to become writers before they realized their mistakes. "Ten years after Wendell went home, I did the same, and his influence figured largely in my decision."[11]

And long before Salatin made headlines with Pollan or in the *New York Times*, Logsdon was including Salatin among the ranks of the "contrary farmers," recommending him and his unconventional methods for raising both chickens and rabbits in *The Contrary Farmer's Invitation to Gardening*.[12] Salatin for his part has repeatedly recommended Logsdon's and Berry's insights in his own books, articles, and blogs.

Salatin includes a 2010 letter from Berry as the preface to his book *The Sheer Ecstasy of Being a Lunatic Farmer*. Salatin had sent Berry an advance copy of the book, and Berry responded to it. In that letter Berry begins by acknowledging that they probably will not agree on all points, but he believes that "the things we would need to discuss are remarkably few." Berry says that he admires this book and commends its "unflagging, exuberant" attention to detail. "It is attention to detail that makes farming an art, and it pleases me that you recognize good farming as a fine art." Berry is then grateful for Salatin's attention to what Berry calls "the formal aspect of farming: [...] the connecting of its various parts so that they sustain one another and become whole." And finally he commends Salatin's insistence upon "the connection of farming to its contexts of ecology [...] and economy." According to Berry, this book of Salatin's is "full of good sense, and when necessary it provides the indispensable wisdom: 'Mark it down, if it smells bad or it's not beautiful, it's not good farming.' Yes indeed."[13]

THREE RESPONSES TO THEOLOGY

For all their similarities, these three farmer-authors also represent very different perspectives on the role of theology in its contribution to farming, ecology, and human flourishing. Wendell Berry's relationship to the Church and to Christian theology has been the subject of a great deal of reflection and debate. It might not be possible to say something uncontroversial about Berry and faith, but perhaps the following points would not be too hotly contested. Berry's written work emerges out of a lifetime of reading Scripture and attempting to understand and apply its meaning. Whether in reflecting on the Sermon on the Mount, the nature of Sabbath, the goodness of creation, or so much more, Berry has been formed by and in conversation with the Christian tradition. Jeff Bilbro has recently argued persuasively that, beginning in about 1979, Berry's rhetoric shifts substantially as he begins speaking explicitly out of the Christian tradition.[14]

Berry, however, is also sharply critical of how many Christians, and the Church, have failed to exercise faithful stewardship of creation. According to Berry Christians have not only condoned but also perpetuated the destruction of the creation. Berry makes this point in both his fiction and nonfiction. Preachers especially do not come across very well in the Port William fiction. As the character Burley Coulter puts it when writing to his nephew Nathan serving during World War II in *A Place on Earth*, the town preacher has "a knack for the Hereafter. He's not much mixed with this world. [...] I wouldn't try to say he ain't right. I do say that some people's knack is for the Here. [...] And surely the talk of a reunion in Heaven is thin comfort to people who need each other here as much as we do."[15] This distinction between a knack for the Here and one for the Hereafter also runs through both Berry's fiction and nonfiction. He is most critical of the Church when it abandons its commitment to the earth and the communities that draw their livelihood from cultivating that earth.

In Logsdon's autobiography *You Can Go Home Again*, he tells the sad story of how, as an adolescent craving attention, he let Sister Monica convince him that he was destined for the priesthood and then talk his family into sending him to a Catholic preparatory seminary in Indiana, far from his home in Ohio. Despite overpowering homesickness, theological doubts, and frequent misbehavior, he stayed in and ultimately graduated from the preparatory seminary, did his novitiate year, took temporary vows, and then went on to a "seminary college." While at the seminary, he spent most of his time working on the seminary's farm and arguing with the priests who ran the school. As his love of farming grew, his already fragile devotion to the priesthood withered and died, taking most of his Catholic faith with it. For some time after this, he pretended to believe so as to not hurt his parents' feelings or alienate himself from his family and friends.[16] In time he would give up the pretense.

Where Berry's criticisms of the Church are usually focused on the fallen nature of people or on the mistaken and arrogant understandings of the "dominion" imperative in Genesis, Logsdon turns his ire first on the "witch-doctor hierarchy"[17] of Catholicism and then on the very idea of a personal God. The last book he published before he died is entitled *Gene Everlasting: A Contrary Farmer's Thoughts on Living Forever*. Like all his books, it is filled with practical wisdom and hard-earned insights about the farm and all its inhabitants and always accompanied by his great wit. But this book is also sad. Here he reflects on his own fight against cancer and how he became convinced "that the various religious beliefs about life after death were just plain ridiculous."[18] Logsdon's vision of living forever is that "our bodies are made of chemicals and organic matter that do not go to any paradise up in the sky somewhere but into a peaceful slumber of humus in old Mother Earth where, in one form or another, they live forever."[19]

Joel Salatin describes himself as a "Christian libertarian environmentalist capitalist lunatic farmer," and he vigorously affirms every adjective in that title. He enthusiastically occupies the space between the antireligious environmentalists and the ecologically apathetic Christians. Salatin thinks he is the only one who is really consistent. He draws attention to the fact that Christians have so often mischaracterized the environmental movement and failed to understand and appreciate the complexities of our common ecology. In speaking about the importance of the microbiological world to soil health, Salatin writes,

> Instead of listening to the nature freaks and foodies who began warning that pesticides, chemical fertilizers, herbicides, and food unpronounceables might be injurious, the religious right branded all these ecology evangelists as pinko commie whacko liberal cultists. But these were the folks who understood how intricate and complex this invisible [microbiological] world is. [...] I realize it's too bad that ... [many] who espouse such views have denied creation, God, and the whole spiritual universe. I get that. It's tragic.
>
> But to ridicule those who have a much deeper respect for the complexity of life just because they don't know the Designer shows a profound immaturity on Christians' part. The world we live in is holy. The life we embrace is holy. My dad used to say, "To us, every bush is a burning bush." When we inculcate in our lives an invisible-world respect and common thoughtfulness, we create the mind-set and the patterns for appreciating that the most real world of all is one we don't even see: heaven, hell, God, Satan, angels, and demons.[20]

For Salatin, ignoring the rich, complex, and all but invisible microbiological world of soil health is like ignoring the invisible world of angels and demons; one does so only at great risk. While Berry and Logsdon are long on the importance of the microbiological world of soil health, you'll not hear them comparing it to the invisible world of angels and demons, except at the most metaphorical or allegorical level.

In his most recent book *The Marvelous Pigness of Pigs: Respecting and Caring for All God's Creation*, Salatin sets up most of the chapters as dichotomies: biological versus mechanical, participation versus abandonment, abundance versus scarcity, integration versus segregation, sun-driven versus earth-driven, beauty versus ugliness, and on and on. In each case Salatin argues that a model dedicated to the faithful stewardship of God's creation and predicated on the holistic use of natural models and patterns is preferable to mechanical, industrial, and chemical attempts to force nature to comply with the designs we place upon it. The latter is the common, conventional way. It is designed with two goals in mind: increase short-term productivity and decrease manual labor. For instance, you can ship cattle to a mammoth concrete feedlot where you quickly fatten the cattle on high-protein grains to a desired sale weight (and then ship the mountains of manure the cattle create somewhere else), or you can slowly grow cattle on a diet of diverse grasses in which you move the cattle daily from one field to the next to ensure the long-term health of both the cattle and soil. One model follows the example of natural grazing that God intended for creation, and one model creates a pattern for quick, human profitability to the ultimate detriment of cattle, earth, and humanity. The contrast is summed up in one of Salatin's final chapters: "Narrow Way vs. Broad Way."

And there one can see three models of how theology relates to agriculture. Theology can play a foundational role in the interpretation of doctrine and implementation of the truths of that doctrine, as it does with Joel Salatin. It can serve as a rich background of moral and religious truth that, when it is not being misused for selfish motives, can inspire and correct both our belief and practice, as it does in Wendell Berry. Or it can become a superfluous abstraction that distracts and misleads one from the essential tasks of caring for the land, as it does in Gene Logsdon.

Each of these three farmer-authors is, in turn, inspiring and maddening. Berry's eloquence, passion, and insight have inspired us and given us a language for how to think and speak about our own farming journey and our place in the world. But as a philosopher, I regularly shake my head at the apparent inconsistencies and contradictions that this passion and insight occasionally produce. He is variously both a libertarian (who wants to be free to do as he pleases) and a communitarian (who privileges the demands that a community places on its members). No author has given us more practical assistance and help in the daily work of the farm than Gene Logsdon. And yet his flights of fancy and hyperbolic disgust for the Church (especially in his fiction) frequently leave one with the impression that "contrariness" has slipped into contempt. And while we are inspired by Salatin's enthusiastic embrace of the close reading of Scripture, his earnestness and enthusiasm sometimes lead one precariously close to some of the very fundamentalist strains of Protestantism that his work otherwise critiques.

I cannot "solve" the problem of theology's relation to agriculture or explain which approach is "correct." When it comes to theological appropriation, I find myself somewhere between Berry and Salatin. But all three of these approaches accurately respond to certain uses, misuses, and abuses of both central theological and agricultural impulses and orientations. There is a lot of nonsense (and even more corruption) that has gone forth under the banner of the Lord and the Lord's work. Specious proof texts and pseudojustifications for everything from slavery to racism to unbridled greed, pleasure, and violence have all appealed to self-indulgent interpretations of Scripture and tradition. The so-called "Dominion theology" and the exhortation to "subdue the earth" found in Genesis 1 has covered a multitude of sins. For all such as this, Logsdon's rejection of such balderdash and hypocrisy seems entirely appropriate.

But of course Logsdon did not merely reject the misuse and the failure to understand and apply the theological insight. He rejected even the possibility of its truth, which is disappointing in so many ways, not the least of which is his own frequent celebration of wonder and mystery and of the hidden life of the gardens and the fields that always surpass our capacity to know all that there is to know about them.

The answer, it seems to me, is most obvious—at least for the farmer. It is found in humility—before the land, before the peaceable kingdom of the barnyard, before the family and the neighbors and the customers with whom and for whom we work, and yes, before the Lord. Every farmer, whether she or he is a Christian or not, knows how easily it is to be humiliated on the farm. In their best moments each of these authors shows the beauty of this humility. And in each of them we also see moments of hubris. The point, it seems to me, is to remember the gift.

In a time when our technological progress and prowess have tempted us to believe that we can extort limitless productivity from our fields and our flocks and our genetically modified crops, we need to return again to the recognition of how we have damaged the gift of good land that has been given to us. We forget at our peril that all that is is finite and must be cared for. It is a gift. How best to care for this gift will be a contested matter. But it will not be surprising if some, like Berry or Salatin, reject the wisdom of the cultured despisers of our age for a more modest, humble, divinely inspired directive. We remain both the steward and the student of creation and its creator, and for some of us, it means that we live and work and have our being in the light of the psalmist's exhortation to remember that "The earth is the Lord's, and the fullness thereof, the world, and all who dwell therein" (Ps 24:1).

DO SWEAT THE SMALL STUFF

A number of years ago, Richard Carlson wrote a popular book entitled *Don't Sweat the Small Stuff... and It's All Small Stuff*. The book became a bestseller and was so popular that it produced its own cottage industry of spin-off products (more books, calendars, notepads, posters, etc). The main point of the book is that we should learn how not to be consumed and paralyzed by much of the minutiae of our lives that constantly threatens to swamp our personal and communal canoes. There is a great deal in Carlson's book that I think is very helpful, and in many respects it's a charming, witty read. However, at times it's also deeply misleading, and in a number of instances, quite simply false.

Especially on the farm, the small stuff can be exceedingly important. Paying attention to the small things can help you avoid big disasters later on. A loose—or worse, broken—fence wire may not seem like a problem, but it won't take long for a sheep or a calf to find the gap in the fence and make its way to the greener pastures on the other side. (Remember, the grass is always greener...) What looks like a small water leak today will be a muddy swamp by next week. And spring winds can turn a loose piece of roof tin on the barn into a skylight in no time at all. (And that now-liberated piece of tin will become a dangerous projectile as it slides, or flies, off of the roof toward the unsuspecting livestock below.)

All barns, sheds, and fences require repairs, and usually that means that excess nails, screws, staples, and bits of wire get left accidentally in the high grass and dirt near where the repair occurred. Baling wire comes off of hay bales and stays in the pasture. It's a little-known fact that there is a magnetic attraction between nails and wire and rubber tractor tires. Therefore, when you drop a nail, pick it up. When you see bits of wire or old screws, put them in your coat pocket.

When you go through a gate, close and latch it. You may think, "I'm coming back this way in just a few minutes, and I'll close it then." This way of thinking leads to sheep on the road. When you turn on a water faucet to fill a trough, stay with it until it's full and then turn off the water. There is no telling how many times we have turned on a hose and while the water was filling up, left to do one quick task and only remembered after we discovered a small pond hours later. We've all done it, and it creates a muddy, wasteful, and expensive mess.

Perhaps some of the most dangerous (and unsuspecting) of the small stuff is bits of string. Every bag of feed is closed with string and a paper tab. You pull on the tab and it releases the string that opens the bag. If you're balancing a fifty-pound bag of feed on a trash can or trough, and especially if you are surrounded by eager animals enthusiastically trying to get to their breakfast, the last thing you are thinking about is that bit of string. But it has got to be picked up and thrown away. On numerous occasions we have had chickens, ducks, turkeys, geese, and guineas get their feet tangled up in these strings. The string gets tighter and tighter, cutting into their flesh and cutting off the circulation. Some of these birds can hardly walk, but they can still fly, and this makes them very hard to catch. Even after you catch the bird and cut the string away, the damage might already be done. We had a duck once that lost a foot because of her string tourniquet. She was not long for this world.

Order and structure are also essential for a well-run farm. (Andrea is laughing hysterically that I would write a sentence like that.) I may be a hypocrite, but that doesn't change the truth of the matter. Every day you use a variety of tools, and each one has to be put away. If not, you won't be able to find it when you need it. And since there are so many different kinds of tools (not to mention an endless variety of sizes and kinds of nails, screws, bolts, etc.), it's essential that you have some order. Otherwise you will spend half your time looking for the tool or the piece of equipment you need and the other half of your time going to the store to buy what you already have but can't find. Putting away a box of wood screws, or a roll of electrical tape, or a pair of needle-nose pliers, may seem like small stuff, but it's actually very important. As Wendell Berry's Old Jack used to say when asked where something was, "Hanging up!"

Order is also important in the sense of the order in which we do tasks. Farm chores need to have an order to them if they are to be done well or even at all. Most of our animal pens open onto a small corral that opens up to the pasture. The sheep sleep in the corral at night. If the pasture grasses are thin and we're supplementing with feed, it's important to feed the sheep first. If you feed the sheep first, then "Zeus" (our large and very strong ram) will be distracted, and he won't run you over when you try to take feed to the pigs or the chickens, which also have pens within that corral. Doing chores in a good order also makes it less likely that you'll forget to do something (or forget to undo it—like turning off the water you turned on). And doing them in order saves time and energy. Once again, it may seem like a small thing whether you remember to get some scratch grains to throw to the chickens on the way to the pigs, but if you forget, then you'll have to go back through two more gates and make a separate trip to the feed barn. And since you're now irritated and in a hurry, that's two more gates that just might get left open.

If by "don't sweat the small stuff," Carlson meant that we should not worry or be anxious about many small matters, then he's mostly correct. But if he means that we should not pay attention and attend to the small stuff, then this often good advice can have dire consequences. Of course, you can't pay attention to everything, and this means that you need to learn what to "sweat"—and what to let go. Most of us just think of this as keeping things in proper perspective and developing the right sorts of habits, especially the habit of attention.

NOT SO HUMBLE,
BUT NEAR TO THE GROUND

We didn't get our first pigs until well into the sixth year of our farming adventure. We had tried our hand at mules and goats and sheep and cattle and all manner of poultry before we ventured into the world of pigs. Like most people, we had a lot of misconceptions about pigs. We were apprehensive of the smell and the mess and the slop. But we should not have been. As William Cobbett says in *Rural Rides*, "A pig in almost every cottage sty! That is the infallible mark of a happy people."[1] And he's right. Pigs are magnificent.

After studying a variety of breeds, we decided that we wanted to raise Berkshires. An old English breed, these medium-sized black hogs have distinctive white "points" (feet, snout, and tail). They have a reputation for being fairly docile, and they don't get to be *too* big. Their distinctive red meat is quite moist due to its high percentage of intermuscular fat. Traditionally pigs were classified as either "lard" or "bacon" breeds, and Berkshires, because of their high fat content, fell into the lard category. Today the demand for hog lard has been almost entirely replaced by vegetable (for cooking) and synthetic (for maintenance) oils. Consequently, breeders have turned the Berkshire pig into a much leaner animal that still maintains its high marbling content, giving the cooked meat a moist, juicy flavor.

Despite their growing popularity in artisan food markets, "Berks" are not that common in Central Texas. In fact, when we started looking for some, we had difficulty finding a pair that fell within our desired constraints of price and geography. In time we found a new registered litter (a "farrow" in pig lingo) in Bowie, Texas, about three hours away from us. Andrea and I drove up to get them the week before Christmas. We got two gilts (young females), put them in a dog crate filled with straw in the back of the truck, and brought them home. We named them "Lucy" and "Peppermint Patty." And we had loads of fun telling friends and family that we had given each other little Berkshire pigs for Christmas.

By late spring we had found a young Berkshire boar that was only an hour and half away. We paid our deposit and waited for him to ween. He was born on Valentine's Day, so we named him "Valentine" because we wanted him to be the romantic type. We had been advised that we shouldn't breed the gilts until they were about ten months old, and if we could find a boar who was a few months younger than the girls, this would make our lives much easier. We could keep them all together at first, since he would not physically be able to mount them until he was about six months old.

Pigs are a bundle of contradictions, and this confusion about pigs has made it into our daily clichés and vocabulary. There really are no compliments that compare one to a pig. We may say that someone is as "strong as an ox" or "busy as a beaver," but when it comes to pigs, it's all negative: "root, hog, or die." We say that someone "sweats like a pig" or "lives in a pigsty." These notions are just wrong. Though pigs have a horrible reputation for being dirty, they are among the very few animals on a farm who will not defecate in their sleeping quarters. And you can't "sweat like a pig," unless of course you do not sweat at all. Pigs do not. In fact, it's because they don't sweat that they like to roll around in cool mud or pools of water in order to cool off. Once they have done so and the mud that had cooled and coated their bodies dries, it falls off, revealing them to be quite clean underneath.

Of course, most of what I'm saying here does require rather substantial qualification. Yes, when the dried mud falls off, they are clean underneath. But until that time, they are a muddy mess. In "Doing It and Getting It Done," I have spoken of the difficulties surrounding staying clean while working with the animals. There is no possibility of staying clean when it comes to working with the pigs. And yes, they will not defecate in their sleeping quarters. But they might decide that the marvelous covered house that you worked so hard to build for them to sleep in would be a better bathroom than a bedroom.

Part of what makes pigs dirty is that it's essential that their pens be kept damp. Pigs are particularly susceptible to various swine respiratory diseases that farmers often call "dust pneumonia." There are many causes of this problem, but they are all aggravated by dry and dusty conditions. One day we showed up at the pigpen only to discover one of the sows in a lethargic state with an intermittent cough. I called our friend Art Hunter, the FFA teacher at Crawford High. Art said that while we should certainly check with the vet, we were probably going to need to give her a shot of penicillin.

The vet confirmed, and the feedstore sold us the penicillin and a VERY large hypodermic needle. Then I had to give this pig a shot into the muscle just behind her shoulder. If you're taking notes, you might want to put down that it's best to hose down the pen AFTER you have caught the pig and given her the shot. So how do you catch a two-hundred-pound, slippery pig in a muddy pen who is running and squealing with two other pigs? For ten minutes my friend Cooper, my son-in-law Drew, and I chased this pig around while I held aloft a very large open hypodermic needle. Slipping and sliding in the newly hosed pen, I almost gave myself this shot of penicillin on several occasions. Fortunately young Valentine came to my rescue. Just when I was about to give up, Valentine cornered Lucy and attempted to mount her. While he wasn't quite old enough to get the job done, she did stand still long enough for me to give her the shot. I'm sure that's not how they teach it in vet school, but it got the job done. And within a few hours, Lucy was back fighting at the trough.

Perhaps the most beloved pig in modern literature is Wilbur from E. B. White's *Charlotte's Web*. It's Wilbur the runt who is saved from the ax, first by Fern and ultimately by Charlotte. At one point while trying to describe Wilbur, Charlotte says, "'Humble' has two meanings. It means 'not proud' and it means 'near to the ground.' That's Wilbur all over. He's not proud and he's near the ground." That might have been true about Wilbur, but it's only half true of the pigs we have had. They may be near to the ground, but they are not humble.

It is not quite right to suggest that pigs are arrogant, but they are exceedingly intelligent, and they are not afraid to let you know that they know all about what's going on. They learn your patterns of behavior and can anticipate your actions even before you're aware of what you're planning to do. They are honest and forthright. When they are happy, their tails curl tightly, and they make grunts and snorts that express deep satisfaction. When they are angry, their tails are straight, and they will tell you all about it in distinctly different tones. Their favorite activity, after eating, is to make a bed in the cool earth, surrounded by soft straw, and take a nap. See? How much smarter can you get than that?

The best book that I have read about pigs is Mark Essig's *Lesser Beasts*. Essig notes that the pig's snout is more like an elephant's trunk than a mere "nose." This remarkable appendage can root around in the toughest ground and yet never lose its "eraser soft" quality. It can distinguish which items are edible and which are not.[2] The pig knows the difference. She just knows.

Essig subtitles his book "A Snout-to-Tail History of the Humble Pig." I think this supposed porcine humility is more an indication of the low esteem in which pigs have been held than any virtuous humility on the part of the pig. Pork is considered "unclean" within both Judaism and Islam, primarily because the omnivoracious pig will eat other unclean items, such as carrion and even excrement. (Essig notes how pigs have been shamefully used to remove even human waste in many cultures over the centuries.) The "humble" pig was often humiliated, kept in tiny unclean quarters and fed scraps and trash. But there is nothing about the pig itself that is very humble. Indeed, pigs are proud.

It's on account of two such proud pigs that I should come clean and mention that there probably was an additional reason we chose Berkshires. Two of the most famous literary hogs happen to be Berkshires as well, and neither of these were very humble. However, since pigs love defying generalization, it's appropriate that one of these literary pigs is noble and good and one of them is evil and malicious. I am speaking of course of those exemplary Berkshires, the Empress of Blandings and Comrade Napoleon, Father of All Animals and Terror of Mankind.

Not So Humble, but Near to the Ground 123

Comrade Napoleon is of course the notorious Berkshire who orchestrates the animal revolution and assumes the dictatorship in George Orwell's *Animal Farm*. Napoleon is the boar whom we love to hate. The Empress of Blandings is Lord Emsworth's prize-winning Berkshire sow. She makes her appearance in a number of the Blandings stories by P. G. Wodehouse. She lives in a marvelous pen not far from the kitchen garden of Blandings Castle. No pigsty this, the Empress's quarters are variously described as a "bijou residence" or her own special "boudoir." In moments of crisis or boredom, Lord Emsworth will usually be found retreating to the serene presence of the Empress and listening for "a sort of gulpy, gurgly, plobby, squishy, wofflesome sound, like a thousand eager men drinking soup in a foreign restaurant."[3]

I can't imagine that our Berkshires will ever have such illustrious careers as either the Empress or Napoleon (and I certainly hope we don't give them the reasons Jones did to rebel), but I do know that we're loving living with and learning from these dear creatures. They may not be humble, but they definitely keep us grounded.

SAVING SPIDERS

Fairly often someone will ask us whether our farm is "organic." This is the litmus test for many hipsters, foodies, and assorted suburban conservation aficionados. Organic farms are pure and natural. Inorganic farms (is there such a thing?) are believed to rely on chemicals and pesticides and who knows what else. Our farm is certainly not "certified" as organic, and in the world of Farmer's Market Puritanism, it's the certification that matters. But it doesn't really matter for us; we're far too small to pursue certification, and what little we do produce is mostly consumed by family and friends. All that notwithstanding, it turns out that we are almost entirely "organic," and, as is often said, we have the weeds to prove it.

"Almost entirely." Two words that might cover a multitude of sins. There are two big exceptions when it comes to our disavowal of chemical pesticides: fire ant killer and wasp spray. I don't even know if the use of these products would count against organic certification, but on no account could we do without them. We buy "flying insect repellent" in bulk, keeping a can in each of the barns and in the garage and in easy access of the porches. Fire ant killer comes in big bags that we spread generously on the mounds.

I do, however, harbor some guilt over this relentless battle with the wasps, the ants, and various assorted venomous spiders. Perhaps if I were more compassionate, I could just let them be. Anyone who has read even a little bit of Iris Murdoch's fiction knows that one of the marks of moral sensitivity is the willingness to look out for the tiniest creatures in our community. Iris's characters, at least the good ones, are forever "saving spiders" and rescuing other insects. In *The Bell* Dora is so focused on liberating a captive butterfly from the train that she forgets Paul's hat and notebook (as well as her own suitcase), leaving them all on the train. (Bringing the notebook and hat was one of reasons for this difficult trip.) Moy saves all manner of insects in *The Green Knight*, and in *The Philosopher's Pupil* Gabriel is described as believing that "the flies which had to be caught and let out of the windows, the wood lice which had to be tenderly liberated into the garden, [and] the spiders which were to be respected in their corners [. . .] had all a life and being all their own."[1] Even troublesome characters, like Charles Arrowby in *The Sea, The Sea*, rescue spiders. Sometimes one character exhorts another to do the saving; in *An Accidental Man* Gracie encourages Ludwig to rescue a struggling fly caught in a glass of milk, and Hilda requires Rupert to save a bumblebee who has fallen into their diminutive swimming pool in *A Fairly Honourable Defeat*.

Perhaps most famously in Murdoch's very first novel, *Under the Net*, Jake Donaghue remarks, "In my experience the spider is the smallest creature whose gaze can be felt."[2] Jake has walked into a dark room and has the sensation of being watched. He seeks out and finds the tiny spectator. Thinking about this line, I have on occasion attempted to experience this alleged gaze of the spider. I have felt the gaze of many other creatures. I know what it feels like to be examined by the mule, the heifer, the ewe, and the ram. And I have felt the gaze of the rattlesnake whose pitch-black eyes draw a bead upon me that is not deterred by the furious rattle. But I have never felt the gaze of a spider. Coming upon a large garden spider in the barn, head high with an elaborate web, I've tried to imagine her as Charlotte, and wondered if she were looking at me. "Some farmer?" Nope. I've thought of Jake and Iris and even Elwyn Brooks himself, but to no avail.

I don't hold a grudge against spiders. In fact, we are always grateful for the large garden spiders that make their webs in the corners of the barn and eat so many of the insects that are harmful to the garden. We don't treat (most) spiders like ants and wasps, but we don't "save" them either. And we are always on the lookout for the Brown Recluse spider, whose sting can be so very nasty, and for scorpions, who love the dark, moist, cool insides of boots and shoes so casually kicked off on the doorstep. Everyone knows that you only have to be stung once to learn the importance of shaking out your boots before you put them on. Last year there was a man on the news who was bitten by a small copperhead that had nestled into the bottom of his shoes. The man spent several days in the hospital before being released. You can bet he checks his shoes from now on.

In the summertime wasp nests are everywhere on our farm, and I have been stung more times than I can count. In our equipment barn, with all its rafters, nooks, and crannies, the wasps, yellow jackets, and dirt daubers are everywhere. Sometimes they sting, but even when they don't, they swarm around you, chase you out of the barn, and sometimes make you trip and fall. Outside of the barn, they will build their nests under eaves, on a fence post, on the ground in the garden, on bushes and picnic tables and benches, or on the back side of a swing. Once after mowing for a couple of hours, hot and sweaty, I went to take a well-earned break on a shady swing suspended between two beautiful pecan trees. I sat down and stretched my arm out along the top edge of the swing, only to be attacked by a swarm of yellow jackets. Half a dozen stings later, my hand was so badly swollen that I couldn't grip a pen.

Fire ants are even worse. Most people in the South, especially in Texas, know all about them. These tiny red ants that will build their mounds almost anywhere are the bane of existence to anyone who wears sandals or pulls weeds or simply takes a log from the woodpile for the firepit. They attack en masse, and their victim immediately knows of their arrival. What's worse, they're fast. Before you know it, they will encompass your feet and ankles and begin making their way up your calf. You can jump, shake, swat, kick your shoe off, and run for the water hose, but you won't get them all. And they will leave their marks. The immediate pain of the stings passes pretty quickly, but the tenderness and itching of what looks like dozens of little white-headed pimples on your feet and ankles lasts for several days. Once while trimming the high grass along a fence row, I stepped on a mound (without my boots) and got more than one hundred stings. My entire left foot and ankle looked like it was going through puberty once again.

These problems, like many others, are often matters best understood and remedied by pace and patience. Usually we're not patient enough to live (in Gene Logsdon's words) "at Nature's pace." In the ideal world, nature will solve many of her own problems. The spiders will eat the wasps, the armadillos will eat the fire ants, and the snakes will eat the rats. Unfortunately the snakes will also eat the chicks, and the armadillos will destroy the garden. There are solutions, but we will only find them when we treat the farm—and the rest of our lives—in the same way that we smoke barbeque. We've got to cook it low and slow and resist the temptation for the quick fix. If we can do that, we might be able to follow Iris's example.

SNAKES AND CHICKS

"Snakes are very shy." This is Mr. Phuti Radiphuti's justification for not killing a cobra that Mma Makutsi found behind the washing basket in the bathroom of their new home. On Phuti's reasoning one should not kill snakes—not even a cobra—because their shyness means that they will avoid people and not come back to where the people are. This theory, of course, turned out to be false—which Phuti would have remembered if he knew his Rikki Tikki Tavi. He would also have known that cobras not only come back; they come back with their mates. Mma Makutsi would not discover this fact until the moment she was about to give birth to their firstborn child, Itumelang Clovis Radiphuti.

I was thinking about Phuti's view when Andrea and I encountered a large rat snake attempting to get into the brooder on the back porch of the new house. We had moved the brooder from the garage at the Cottage because one of the hens (our only Bantam) had just hatched seven chicks in a wagon kept in the barn. Knowing that there was virtually no chance that these chicks would make it on their own (when they fell out of the wagon, there would be no way for them to get back or for Mama to protect them), we moved the brooder to the back porch of the new house, collected the chicks, and installed them in their new home.

The chicks had been there for a couple of days when we found the snake. It was ten thirty at night when I went out to check on the chicks before turning in. He was about four feet long, and he had climbed all the way to the top and was trying to work his way past the hardware cloth lid on the brooder. If he had climbed from the front of the brooder, he probably could have squeezed his way in through the crack where the shaving pan comes in and out.

Before I called Andrea to come see our new visitor, I had already begun to wonder what we should do with him. Rat snakes are not venomous, and they serve an important function on the farm, where rats and mice might congregate in the presence of spilled chicken feed. We do not deliberate about the killing of rattlesnakes, but rat snakes are a more complicated issue. But however valuable he might be, now that he'd learned where the baby chicks were, he couldn't simply be moved into the pasture away from the house. He would come back, and rat snakes can be aggressive. He would have to go.

But how? All of our hoes and rakes and the pistol with rat shot were at the Cottage and in the barn. The only guns we had at the new house were far too powerful to use on this snake. When Andrea, who was already in her pajamas, came to see him on the porch, I told her to watch him and keep him from getting inside the brooder while I went to the Cottage to get the pistol. "Are you crazy?" she said. "I'll go get the gun. You keep him from getting in the brooder!"

While she went to the Cottage in her pajamas and Hunter boots, I found a couple of long paint roller sticks. The hooked tips looked sort of like the tools those professional snake handlers use on Animal Planet. "This will be easy," I thought. "I'll pick him up with this roller and move him to the yard where I can shoot him." When Andrea got back, I checked the gun to see that it was loaded and asked her to hold it while I got the snake. "Do you want the safety on or off?" "ON—of course."

My two paint rollers were like giant chopsticks. Picking up the snake with the sticks was harder than I imagined because as soon as I went for him he started moving. When I finally got him off the brooder and attempted to flip him into the yard, I didn't realize that he had wrapped part of his body around the stick. So when I went to toss him into the yard, he didn't go into the yard. I inadvertently threw him on Andrea.

All at once several things happened. As she screamed, the snake fell to the ground and began slithering around and between her feet. She couldn't seem to get away from it, frantically dancing around in her pajamas and boots while waving the pistol aloft. I was still going after the snake with my oversized chopsticks, sometimes hitting the snake and occasionally hitting Andrea as well. I was not doing this very well because I was laughing so hard. It took Andrea longer to find the humor in this situation. The snake retreated under a chair and then under a table. I got him off the porch and into the yard, but before I could get the gun from Andrea, he had made his way back on the porch and was headed for the back door.

We finally cornered him under a bench and then he disappeared. Where did he go? Pulling out a small trash can that we use to store the dog food, we saw him coiled behind the trash can. He was of course no match for the rat shot, and after I dispatched him to the pasture where the buzzards would find him in the morning, we both finally had a good laugh.

Once inside we discovered that the commotion on the porch had been so traumatic for the nine-year-old Cocker Spaniel, that she ran upstairs to hide, apparently peeing as she went up the stairs. There were small pools of dog pee on several steps of our new staircase, and it was even dripping off the edge in a couple of places. It took us several minutes to clean it all up.

As Andrea's Aunt Kay likes to say, "You can't make this stuff up."

But were we right to kill the rat snake? Will we be overrun by rats and mice? Of course, we cannot compare a rat snake to a cobra. Phuti's cobra (and his mate) did come back, but that story ended with Phuti and Grace being grateful for the snake—since he, before his death from mostly natural causes, accomplished what the couple themselves never could have. And Phuti, or rather Mealies, the snake-killing dog Phuti borrowed from a friend, did attempt to kill the snake, even if he got the wrong one on the first try. But at least the chicks are safe—for now.

TOLSTOY AND PAHOM

Leo Tolstoy's marvelous short story "How Much Land Does a Man Need?" is about a peasant named Pahom who develops an unquenchable thirst for ever more land. He starts with nothing and naively believes that if only he could acquire a little more land, he will solve all of his problems. At first he only wants a small piece of land that he can call his own. Then he needs a little more, and then he needs better land. In time he desires both better land and more land, and of course in the end his avarice kills him.

When we bought the first ten acres of our farm, the surveyors had placed stakes at the back corners of the property. The stakes were in the middle of a large pasture with few natural landmarks. I distinctly remember the first time we walked all the way to the corners of *our land*. All of this was ours. This was a huge tract of land—ten whole acres (slightly more, actually). Looking back to the house and road, how small and far away it all seemed.

Before we could afford a fence, we set tall corner posts in concrete and painted the tops red so that we could always find them. The hill crests in the middle of our back pasture, and if the grass were high, visitors might not even be able to see the corners. "How far does your property go back?" "All the way to those corner posts." "Where? I can't see them. Oh, wow. All of that?" "Yep. All of that." Even after we put in a perimeter fence, it still seemed very large. We planted pecan trees along the back fence and had to haul nine five-gallon buckets of water all the way to the back of the property just to keep them alive. We were a long way from the house, or so we thought.

Over the next few years, that huge tract of land began to shrink. We dug a pond and put in a gravel road. The cattle kept the grass short, and we parked a couple of trailers back there. The back pasture was becoming less of an undiscovered country. After years of planning, we began building a new house on the far side of the pond in the fifth year that we owned the farm. Once the new house began going up, the property really became small. The house had to be fenced off from the cattle, so as the herd began to grow, their grazing land began to shrink. And as we stood on the front porch of our unfinished house and looked beyond our fences, it became very clear that we needed more land.

We had long dreamed of owning the land immediately to the west of our property. There were two really good trees over there, and we had always feared that someone would buy that land and build a house, blocking our view of the sunset. Of course it would be an ugly house, built close to our property line, and they would probably play loud rock music and root for the Aggies. This would be intolerable.

In time we were able to buy the seventeen acres immediately to our west. When we walked to the far corners of the new land, even the new house looked tiny. I took a picture of Andrea in the field looking toward the house, and it reminded us of Wyeth's *Christina's World*. And the original farmhouse cottage, hidden by trees in the corner of our twenty-seven-acre farm, was all but invisible.

Unfortunately it didn't take long before we began looking around for more land. One neighbor said that such and such a piece of property would soon come on the market, and another had bought the piece just down the road for a good price. And if we were to increase the size of our herd . . .

How much land does a man need? We forget about Pahom at our peril.

THE COW IN THE PARKING LOT

There is a popular book by Leonard Scheff and Susan Edmiston entitled *The Cow in the Parking Lot: A Zen Approach to Overcoming Anger*. The title of the book comes from a contemporary Zen parable. In the parable one imagines two scenarios. In the first scenario, after driving around a shopping mall parking lot for ten minutes looking for a space, you finally find a spot from which another car is leaving. You wait patiently for the car to depart (with your blinker on), and then at the last second, some guy in a jeep swerves in and steals "your" spot. You honk, but the driver smirks and gives you the finger. You are seething with (justified?) anger. The second scenario presents the same scene, but instead of the arrogant young man, a cow walks over into the space and lies down in the middle of it. You honk, but she looks up unaffected and doesn't move. Instead of being angry, you are bemused.

The authors use this parable on several occasions in the book to illustrate how one can overcome anger by learning to recognize that our anger begins and ends with ourselves, and that by changing our perspective, we have the chance to live happier and more fulfilling lives. There is much about this book that I like, and some of the authors' strategies and insights are really quite helpful. But I must say, they do not seem to understand cows in the parking lot at all.

In "Ockham, Iris, and the Show Cattle," I have recounted what I called the most terrifying hour of my life: that morning at the county livestock show when, just as I was leading one of our heifers out of the trailer, a large truck drove by and honked or backfired or made some loud, unexpected noise. At that exact moment, the startled heifer slipped on some fresh manure on the concrete, fell, got spooked, ripped the lead line out of my hand, and spent the next hour running around the parking lot. At one point she was dangerously close to getting out of the parking lot and onto a busy four-lane commercial street. At every point in this horrible hour, she was in close proximity to well-meaning children and teenagers who were trying to "help" catch her. And at any point she could have trampled any of them. When we finally caught her and trapped her in the trailer, she was mad as hell. She was a beautiful and expensive show heifer, but we never took her to another livestock show.

Scheff and Edmiston are exactly right that in moments of anger or frustration, one needs a change of perspective. But the perspective of the farmer or rancher, or anyone who lives in close proximity to livestock, upon encountering a loose cow in a mall parking lot would be anything but bemusement. In this situation one can only ask, "What the hell is this cow doing in the parking lot?" There are no good answers to this question.

One needn't have had my horrible experience at the county show to see the issues here. If there were a real cow loose in the parking lot, there's every chance that one's anger, as well as one's fear and frustration, would be magnified rather than diminished. The questions come pouring out, and they are not abstract, theoretical, or conducive to Buddhist contemplation. Whose cow is this? Is she okay? Who left the gate open? How did she get here? Are there other cows loose as well? How are we going to catch her? IF we can catch her, how are we going to hold her? (Remember, depending on her breed and age, she probably weighs between eight and twelve hundred pounds. And twelve hundred pounds goes wherever twelve hundred pounds wants to.)

And this is to say nothing of it being *your* cow in the parking lot. How much money is invested in her? What's going to happen if she gets scared and hurts someone? The likelihood of *her* getting hurt is very high. How long is it going to take you and your family to corral her, get her into a safe spot, and wait for someone to come with a trailer? And all of this is assuming that she's not terrified being in a concrete wilderness surrounded by honking horns, impatient shoppers, and eighteen-wheelers making deliveries, none of whom are accustomed to seeing a half-ton of bovine beauty taking up a parking space.

And what if it's not a she at all? What if it's a bull in the parking lot? Well, now we're getting somewhere. The presence of a mature bull in a parking lot presents the possibility of real danger. There's an old joke about the sign on the pasture gate: "Entrance to the pasture is free, but the bull will charge."

Maybe the situation is not one of danger but merely one of inconvenience. What if you pulled into that spot or one near it, and stepped out of the car and into a fresh cow patty? How would that change your trip to the mall? Whether we're talking about sandals, sneakers, or blue suede shoes, we've definitely moved to a new level in anger management and probably one that transcends how we felt about the guy in the jeep.

In the parable (and the book), one is able to overcome one's anger because the jackass in the jeep (I'm mixing my farm animal metaphors) has been replaced by a peaceful and pastoral cow. This cow is not threatened; therefore, she behaves. What happened before and what happens next to the cow is of no consequence. But what kind of cow would actually act this way in a busy parking lot? Only an imaginary one.

My point here is not to complain about cows in the parking lot or to diminish Buddhist strategies for overcoming paralyzing emotions. Rather, I am reminded again not only about the difficulties of reconciling abstract theories and concrete (!!!) practices but also about the necessity of attending to that which is really there and understanding how I should respond.

On the first point, I hear yet again, "Physician, heal thyself." I am often the one whose beautiful theory doesn't stand up to the scrutiny of experience. Philosopher-farmers are frequently long on philosophy and short on farming experience. And the persuasiveness of my theories will be closely connected to how well they are informed by a world that is always greater and more complex than I had imagined.

The second point follows closely on the first. It is helpful for me to learn to gain new perspectives and to realize the extent to which I am responsible for my emotions (especially anger) and particularly the ways in which my anger is more a product of my own unrealized expectations than most disappointments encountered in reality. But I can't overcome my anger by merely imagining a world that is different from the one that I hoped for or desired or must now deal with. I must attend to the world that is really there. Hopefully I'm learning to attend to it with patience and humility, recognizing that if I let the loss of a parking spot ruin my day, I've got some bigger problems I need to address. The jackasses will always be there. Occasionally there will be cows—and on those days, things will get a little bit more complicated.

BACK TO THE ROUGH GROUND

THE CONSOLATIONS OF *TECHNE*

One of the most prevalent themes in these reflections is my own recurring failures to know how to do this or that task on the farm. Whether it is birthing lambs or training heifers, our farm experience has largely been a matter of learning "how to do" some new agricultural endeavor. The story of our slow education as farmers is the story of how we are attempting to acquire the knowledge and the skills to be successful at farming. This knowledge and skill is what the ancients called *techne*. *Techne* is one of the fundamental intellectual virtues or ways of knowing. It is of course the root word from which we get words like technical, technique, and technology.

On the farm, one of our ongoing questions is the matter of how we use technology in the most appropriate way. I cannot of course frame this question in the form of *whether* it is appropriate to use technology; we are always using some kind of technology. I may not be putting ammonium nitrate on our fields, but I do love the pitchfork, and I keep the chain saw oiled and ready to go. Wendell Berry quite famously swore off the use of computers, but (as he noted) he happily uses the pencil and the typewriter. So it is with all of us. Technology in the modern world is almost always a question of *which*, not *if*. But which kinds of technology we use matters a great deal. This, however, is not the question I am primarily concerned with at the moment.

I am concerned with the role that technology plays in how we *understand* and *desire* what kind of lives we want to lead. It is quite common for many (perhaps most) people to assume that whatever problem we currently have can and should be solved by some new technology. If you can imagine it, there is someone working on "an app for that." And this is as much the case on the farm as it is in the city. This notion that technology can save us is sometimes called "technological messianism."

The idea of technology as messiah is deeply consoling to the modern mind. And given the successes of modern technology, it does not seem too far-fetched. From the sublime achievements of modern medicine and the unimaginable feats in global transportation to the gloriously mundane character of cell phones, air conditioning, and the Igloo cooler, technology has a pretty impressive track record. If most people think that technology can save us, it might not be an unreasonable belief. Most of us are not only deeply consoled, we are excited by the possibilities of technology.

And this is clearer nowhere than on the farm. Most modern farmers use levels of technology on a daily basis that city dwellers might have a hard time imagining. Sure, New Yorkers might take the subway or an Uber to their office on the twenty-first floor where they have video conferences with colleagues in Europe or Asia. But can they imagine a half-million-dollar tractor guided by GPS, which not only drives itself but can place seeds within a three-inch variance of perfectly straight lines of rows that go on for hundreds and hundreds of miles? Or a combine harvester, now more than a century old, which quite literally combines the work of three historically separate activities (reaping, threshing, and winnowing) into one single process? Or the cultivation of transgenic (a.k.a. genetically modified) crops that are not only resistant to herbicides but can be shipped all over the world without spoiling? Modern farm life is saturated with the consolations and possibilities of *techne*.

But *techne* is not without its difficulties. As Boethius taught us long ago, "Good fortune corrupts and bad fortune instructs," and in many ways we have been corrupted by the good fortune and the promises that *techne* regularly makes to us. Aristotle knew that *techne* works best (or at all) when it works in concert with the other virtues. Of course, it is not just *techne*. All of the virtues require the cultivation of the other virtues to work well (or at all). As G. K. Chesterton famously pointed out, "The modern world is full of the old Christian virtues gone mad. The virtues have gone mad because they have been isolated from each other and are wandering alone."[1]

This essay is not a diatribe against *techne* itself. *Techne* is an essential intellectual virtue without which one will not flourish or be happy, even and especially on the farm. But true flourishing or happiness requires us to know when *techne* is required and when it is insufficient for the task at hand. Primarily this essay is about learning what kind of life I should be desiring and learning how to evaluate the claims made by a world that wants to solve my problems with ever more technology. Usually this means convincing me to buy and use a new tool.

Aristotle knew that acquiring *techne* was an essential dimension of the good life. But it was never adequate on its own. To work well, it needs the other virtues, especially the virtue of *phronesis* (prudence or practical judgment). These two are often and easily confused. Learning how to operate a chain saw will help me cut down a tree, but it will not help me decide which trees need to be felled for the good of the farm. I fear that all too often, when we glorify *techne* and assume that it will solve all our problems, we also make another mistake. We confuse the "Nice" for the "Good," or to be more specific, we assume that a Nice life filled with technological comforts is the Good Life. The Nice is a false good that perpetuates a blinding, consoling fantasy, and this is one of the principal dangers of our enthrallment with *techne*. On my reading of these matters, learning to tell the difference between the Nice and the Good (and its relation to *techne*) is essential for true flourishing and happiness.

In order to make my argument, I need to bring in some rather disparate voices, most importantly Aristotle, Iris Murdoch, and the philosophers Ludwig Wittgenstein and Hans-Georg Gadamer. Chief among my debts is the marvelous 1993 volume by Joseph Dunne, *Back to the Rough Ground: Practical Judgment and the Lure of Technique*. Dunne's title (and my own), of course, comes from Wittgenstein's aphorism, "Back to the Rough Ground!"[2]

My own conclusion is indebted to Dunne's. We learn to resist the false promises of *techne* through the cultivation of the other virtues, especially *phronesis* (prudence or practical judgment). It may seem odd at first to focus on *phronesis* on the farm since the virtue of *phronesis* mostly applies to learning how to work with people, and *techne* is reserved for those skills that are needed for work with inanimate objects and nonrational animals, both of which abound on the farm. But *phronesis* does not make promises, and unlike *techne* it is very rarely consoling. *Techne* is designed to make things easier. The prudence that is *phronesis* slows us down by making us think again. *Phronesis* is in many respects a "rough ground," a place of friction and contact. This rough ground is hard to till because of the rocks and the thin soil. And yet this rough ground not only enables us to recognize when misdirected *techne* offers false consolation (just buying another tool is not necessarily going to solve my problem), it also equips us to distinguish between the allure of the Nice and the cultivation of the difficult Good.

Technological progress in agriculture bears a resemblance to Shakespeare's great tragic figure, Macbeth. Macbeth is a Scottish lord, known for his courage and his loyalty to the king. But at the beginning of the play, the witches offer a prophecy to Macbeth that changes dramatically how he understands himself and his aspirations. Similarly, the dramatic technological changes that occurred in agriculture in the last century offered visions of productivity and efficiency that could not have been imagined in previous centuries. And yet, like Macbeth, to accomplish these new aspirations, noble lies must be told and great men (and families) will inevitably be sacrificed. By the end of the play, Macbeth has ascended to the throne (he's acquired what he never could have dreamed of previously) but only through the destruction and devastation of Scotland as the characters know it. In the final scenes, we see Macbeth as a character who has accomplished his reckless ambitions and gotten what he wanted, but in the process, lost everything he had. And the world and the life that has been created in the process seem to be nothing but "a tale told by an idiot, full of sound and fury, signifying nothing."

"A PICTURE HELD US CAPTIVE"

In *Philosophical Investigations* Wittgenstein asserts, "a picture held us captive." He is speaking about language. He goes on to note, "And we could not get outside it for it lay in our language and language seemed to repeat it to us inexorably."³ Making the necessary translations, it seems clear to me that modern industrial societies are also held captive by a picture; we are enslaved to a world in which we assume that most problems can be solved by either a technological solution or, at the very least, a technique that can be applied to the situation at hand.

Our phones wake us up to the tune of our favorite song. They tell us how cold it is outside, how many calories we have burned on our morning run, and which route to work has the least traffic. We pick up the coffee and muffin we ordered through an online app on our way to a job where we work on a computer that makes millions of calculations while allowing us to create, to communicate, and to search for information and purchase products from around the globe. Everyday there is a new app or website that will help me accomplish a task, chronicle my life through art and music, or purchase something with ever greater ease.

And this is to say nothing of the entire realm of health care. Though we are deeply grateful for the many thousands of medicines and procedures that daily save lives and alleviate pain, increasingly we assume that any difficulty we have can be solved through medication or a medical procedure. Does my middle-aged body no longer resemble the body of a twenty-something? There are procedures to change that. Are your children distracted or disorderly? Perhaps they need a medication. Am I overweight, tired all the time, unable to concentrate or control my appetites and urges? Perhaps I have a syndrome or a "condition" for which there is medical remedy. And if there is no medical remedy today, perhaps a pharmaceutical company will create one tomorrow.

Many of these things are good in and of themselves. And using an alarm clock function on one's phone is perhaps not substantially different from an old-fashioned alarm clock or what we used to call a "clock radio." (Except one costs $15 and the other $600 to $1,000.) The problem is with the world of expectation and entitlement that the possession of that technology and the appeal to its expertise entails. In his essay "Diagnosing the Modern Malaise," Walker Percy wrote, "When something goes wrong in our technological environment, if something needs fixing, whether it is one's car or one's intestinal tract, we have reason to believe that 'they' can fix it, 'they' being the appropriate specialist."[4] We imagine that technology and its culture of experts can give us a perfect world.

Wittgenstein continues, "For it lay in our language and language seemed to repeat it to us inexorably." "Efficient," "useful," "practical"—these words now carry connotations against which it is almost impossible to argue. These characteristics are thought to be intrinsically positive. Phrases like the "value of inefficiency" are simply incoherent to most people. "What could possibly be good about inefficiency? What could that even mean? If some bit of knowledge is not useful, what good is it?" When I question these kinds of assumptions with my students, they invariably think that I'm trying to make a point through exaggeration.

"A picture held us captive." *Captive?* One might reasonably respond, "Surely this is an exaggeration. Is the consolation of *techne* really such a problem that one would describe it as a kind of prison?" Captivity and enslavement metaphors have a long and varied history in philosophy, but it does seem to me that Wittgenstein's language is appropriate, particularly with regard to the allure of this kind of consolation. It seems to me that the consolations of *techne* occur in a variety of forms, and many of them are indeed pernicious.

The technological messianism of our general culture is exemplified by our desire for ever more sophisticated technologies to organize and (here is the key word) *manage* our affairs and our world. It is not merely the existence of these wonderful gadgets (no small number of which may be found in my own briefcase); it is the delusory fantasy that many of us entertain that, through technology, we will finally be able to overcome all the challenges we face. What was that quotation I was trying to remember? Surely I need to be able to access all of Plato, Aristotle, Augustine, Shakespeare, Dickens, and Dostoevsky on my Kindle wherever I travel. If I am not able to send and receive text messages and phone calls, check my email and social media accounts, and surf the web for everything from news updates to movie schedules, how can I reasonably operate in the modern world? If I do not have the "Find a Starbucks" app on my phone, how will I survive in Honolulu?

Technological messianism is only one species of the consolations of *techne*, however. I am much more concerned about the menagerie of beliefs that arises from our overdependence on technology, and it is these beliefs that are the consolations of *techne*. Take, for instance, the ways in which the easy access to information can corrupt us. When I can always look it up, I have no reason to learn it. As Jacques Barzun noted in *The Culture We Deserve* (long before the revolution of "handheld" technology):

> So much compiling and disseminating of data in small bundles is, among other things, an orgy of self-consciousness. We seem to live mainly in order to see how we live, and this habit brings on what might be called the externalizing of knowledge; with every new manual there is less need for its internal, visceral presence. The owner or user feels confident that he possesses its contents—there they are, in handy form on the handy shelf. [. . .] The master of so much packaged stuff has less need to grasp context or meaning than his forebears: he can always look it up. His active memory is otherwise engaged anyway, full of the arbitrary names, initials, and code figures essential to carrying on daily life. He can be vague about the rest: he can always check it out.[5]

The same goes for my relationships with my students or colleagues. Concerning students, to combat plagiarism, must I require all of my students to turn their essays into a phrase-checking database that will tell me what percentage of their work is original? How is my relationship to them, to their work, to their futures, transformed through the mediation of *techne*? Have I even actually accomplished my goal of detecting plagiarism? If so, for how long? When a university begins to *require* that all papers use "Turnitin.com," what does this do to our communal culture? In what subtle ways do I now believe that what I *really need* to solve this age-old problem is a technological solution?

Similar transformations occur in my relationships with my friends. Notice how language is transformed. Words such as "friend," "chat," "text," and hundreds more all have new meanings. This is not a complaint against "meaning as use." It is the recognition of how technologically mediated engagement becomes the new standard. *Convenience becomes the standard*. Just as one cannot argue against efficiency, one cannot argue for inconvenience. But the problem is far greater than simply convenience or inconvenience.

Certain kinds of technology promise the illusion that we can perfect human experience. To take one trivial example, the use of "instant replay" and "official reviews" have almost ruined the experience of professional and college sports. Any of us who have played or watched sports we love know the anguish and the frustration of a "bad call." But when the game is slowed to a snail's pace in order to get it "perfect" (and even then they miss a lot), it takes the joy out of the game. In the quest for technological perfection, it ceases to be play at all.

We have appealed to *techne* to satisfy our cultural obsession with safety and security. Schools become fortresses. Public transportation becomes an occasion for invasive full-body scans. The FDA requires sanitation standards that all but destroy small farms. All of these practices are "perfected" with the best of intentions. But each in its own way fails to remember that the world is not safe! More than thirty years ago, Wendell Berry was making this argument with regard to human sexuality. "And in the midst of this acid rainfall of predation and recrimination, we presume to teach our young people that sex can be made 'safe'—by the use inevitably of purchased drugs and devices. What a lie! Sex was never safe, and it is less safe now than it has ever been."[6] And what Berry says about sex is true about large portions of the rest of the world as well.

We see the lure of *techne* in our academic disciplines. Greater precision and sophistication in analysis always calls for more exacting, "technical" vocabularies and methodologies. But there is a way in which we can become so comfortable with our methodologies that we fail to recognize or exercise the necessary hermeneutical dimensions to our work. And as Hans-Georg Gadamer notes, hermeneutics must always rely on *phronesis* rather than *techne*. In my own discipline, philosophy, one sees the glorification of *techne* at every turn. Method and procedure reign supreme.

The analysis or critique of work principally becomes the analysis of whether proper procedure was followed. It is a confusion of "validity" with "truth," and in such an environment, *techne* becomes the new *sophia*, and thus what was traditionally the "love of wisdom" becomes the "love of *techne*."

I believe that this is the underlining story of the tragedy of industrial agriculture. For farming families that were always a drought away from foreclosure, the desire for increased productivity was always present. New "labor-saving machines" allowed more work to be done. But to pay the debt on the machines, more land had to be farmed and even bigger (and more expensive) machines had to be purchased. To maximize productivity, chemicals and short-term fertilizers were developed. This became the new cycle. One must "get big or get out." Gadamer was not explicitly referring to farming, but his insight rings true nonetheless: we can become so comfortable with our methodologies that we fail to recognize the necessary hermeneutical dimensions to our work. Where did it all go wrong? How should *techne* be properly understood?

TECHNE AND *PHRONESIS* IN ARISTOTLE'S *ETHICS*

The most important classical formulation of the distinction between *techne* and *phronesis* is found in Book VI of Aristotle's *Nicomachean Ethics*, in which he turns to the examination of the intellectual virtues. The five primary intellectual virtues he discusses are *episteme* (scientific knowledge), *techne* (art or skill), *phronesis* (practical judgment), *nous* (knowledge of first principles), and *sophia* (wisdom). For our purposes, we need to return to the relationship between *techne* and *phronesis*.

For Aristotle, *techne* is that intellectual virtue exemplified by technical skill, craftsmanship, or artistic ability. (Remember, for Aristotle virtues are dispositions rather than feelings or capacities, so the virtue of *techne* is not the skill itself; it is the reasoned productive state that produces the skill or the ability.) *Techne* is a kind of "know-how" that enables one to know how to build a house, train a dog, paint a picture, or pay bills on a computer. Lest it seem that *techne* is reserved only for mundane activities, it is also the kind of knowledge that enables one to do differential calculus, perform thoracic surgery, or break a colt or a steer. It is also the necessary (though perhaps not sufficient) condition for writing poetry in a second language, dancing the tango, or playing Bach's cello suites.

Phronesis, on the other hand, is usually translated as "practical judgment," "prudence," or "practical wisdom." It is that insight into the world of human affairs that rightly orders action. Aristotle contrasts it with *episteme* (scientific knowledge) because *phronesis* equips one with the knowledge to respond to variable causes (things that are not necessarily the case) and with *techne* because the goods of human affairs cannot be ordered in the ways that inanimate objects or nonrational animals can be ordered. According to Aristotle, *phronesis* is the most important intellectual virtue for political science and life in community because it enables one to understand the variable ways in which human beings relate particular facts to generalized principles. With both *episteme* and *techne* there is a necessary relation of fact to principle.

Hans-Georg Gadamer takes up the theme of *techne* and *phronesis* in his great work *Truth and Method*. It is perhaps not too much of a stretch to describe the most important work in philosophical hermeneutics as an extended reflection on the distinction between *techne* and *phronesis*. One might render the thesis of Gadamer's book that *Method is no guarantor of Truth*; indeed, Gadamer's principal aim seems to be to disabuse his readers of the commonplace notion that somehow Method just is Truth. Gadamer rejects this notion. For our purposes it is important to recognize that Method is a kind of *techne*.

What follows from this distinction between *techne* and *phronesis*? It means that one cannot repair one's marriage in the same way that one repairs one's computer. There is no single set of rules or skills or formulae that may be used to solve the problems that vex human community. Despite the fact that the shelves at Barnes and Noble sag from the weight of all the books that promise to teach us how to solve our interpersonal relationship problems, overcome our dysfunctional families, or win friends and influence people, most of these volumes mistakenly offer the promise of a *techne* to accomplish what only a difficult *phronesis* can achieve.

But what about the farm? The farm seems to be a place filled with inanimate objects and nonrational animals. Would *techne* not be the most important intellectual virtue in this setting? Whether you are using a hoe to weed the garden, a posthole digger to build a fence, or a rope to catch the calf, are these not all appropriate tasks for which *techne* is designed?

The short answer to all of these questions is, "yes, these are jobs for *techne*." And if you do not have adequate agricultural *techne*, you are not going to get very much done on the farm. *Techne* answers many of the "how-to" questions that arise on the farm.

Techne answers the question "how should we be doing this task?" But *techne* is rarely adequate to answer the question "what should we be doing?" *Techne* is not equipped to answer questions about what goals we have for the farm or why we would want to live this way to begin with. If you do not have to live a life shoveling shit out of the barn, why would you want to? *Techne* cannot answer this question.

THE "LURE OF TECHNIQUE" AND THE NICE AND THE GOOD

Dunne's objective in *Back to the Rough Ground* had been to explore the way certain educational techniques ("the behavioral objectives model") that were designed to achieve greater education were in fact undermining the real goals they had been inspired to accomplish. He notes that the "lure of technique" offers a false promise to educators that they need only to learn a certain technique or follow a special set of rules or employ a new procedure in order to achieve all their educational dreams.

It is easy to see how the "lure of *techne*" can corrupt the farmer with the false promises of "the Nice." Certain chemical fertilizers might give me a bumper crop for a few years (wouldn't that be Nice?), but what damage is being done to the long-term health of the soil and the fields (is it really Good for us)? Often my reckless pursuit of expanded productivity comes at the cost of diminishing the very thing (the soil) which will make my life as a farmer possible. To keep from harming the soil I may have to refrain from using certain tempting fertilizers. I may have to seek alternative ways of restoring and rejuvenating the topsoil. This will be difficult. But it might be Good in the long run.

Iris Murdoch does a good job of explaining the differences between the Nice and the Good and how it applies to our lives. Over and over again, throughout both her philosophy and her fiction Murdoch juxtaposes the "Nice" with the "Good." *The Nice and the Good* is in fact the title of one of her novels, but the tension and the difference is found throughout her work. For Murdoch, "False love moves to false good. False love embraces false death."[7] The "false good" that most often tempts Iris's characters is the lure of the Nice and the temptation to confuse it with the Good. The Nice is the cozy, tidy world without muddle, a world of easy love and convenient relationships, an ideal world in which problems disappear. This is the Nice, and it is a consolation to those who think they have it and an all-encompassing temptation for those who do not. In contrast, "the Good" is neither easy nor convenient. It is always difficult and it resists easy definition. It is "the magnetic centre towards which love naturally moves."[8]

I need to be very careful here, because in her philosophical works, *techne* plays an important role in Murdoch's own thinking about the nature of art and its relation to philosophy and to truth. (She notes, "Beauty and the *technai* are, to use Plato's image, the text written in large letters. The concept Good itself is the much harder to discern but essentially similar text written in small letters."[9]) And I do not want to suggest that Murdoch principally thinks that the Nice is a *techne*. However, there are two ways for her in which *techne* contributes to the Nice.

First, professional philosophy has often reduced traditional philosophy to *techne*. As practiced in the Oxford analytical establishment (within which she was both student and professor), professional philosophy is principally directed toward conceptual clarification through propositional analysis. Murdoch claimed to Bryan Magee that "Literature . . . does many things, and philosophy does one thing."[10] On another occasion she noted that "philosophy is fantastically difficult and I think those who attempt to write it would probably agree that there are very few moments when they rise to the level of real philosophy.... I don't think that I have done it in more than a few pages in all the stuff that I have written about philosophy."[11] This is not false modesty on Murdoch's part. As she said to John Haffenden in 1985, "If you're not doing philosophy pretty well you're not doing it at all."[12] Murdoch both idealized philosophy and was frustrated by it, in no small part it seems to me, because of this tendency toward *techne* as the standard of excellence for philosophy.

Second (and of lesser importance, I think), there are indeed some characters in her novels who might be described as pursuing *techne* as a way of achieving the Nice or of finding their consolation in this *techne*. Such examples include Charles Arrowby in *The Sea, The Sea*, Thomas McCaskerville in *The Good Apprentice*, and Hilary Burde in *A Word Child*, among others. On the whole, however, this is not her primary focus. For Murdoch the Nice is a false Good. It is a blinding and consoling fantasy and a product of one's failure to pay attention to reality or "how things really are." My argument moves in the other direction.

I want to suggest that the pursuit of *techne* and its consolations are *for us* a Nice that must be distinguished from the Good. These consolations are a false Good that blinds us to how things really are and corrupts our attempts to be formed by both the world around us and our communities. We are (illegitimately) consoled by our appeal to utility and efficiency, and our clinging to these consolations becomes for us the new Nice.

The more important point from Murdoch's perspective is that *techne* can never satisfy the demands of the moral life, and it is this point that a philosopher-farmer who quests after the new Nice and is consoled by the fantasies of *techne* must learn again and again. What does Murdoch have in mind here? The *techne* that she has in mind is that of modern scientific methodologies and assumptions (for instance, some forms of psychology or psychologically driven philosophical approaches) that pretend to understand the human being so well as to overcome the essential variability of the human person and the human predicament. (It is important for us to remember that this essential variability in human goods is what demands *phronesis* rather than *techne* for Aristotle.) She writes,

But why should some unspecified psychoanalyst be the measure of all things? [. . .] The notion of an "ideal analysis" is a misleading one. There is no existing series the extension of which could lead to such an ideal. This is a *moral* question; and what is at stake here is the liberation of morality, and of philosophy as a study of human nature, from the domination of science: or rather from the domination of inexact ideas of science which haunt philosophers and other thinkers.[13]

Murdoch is clear that her work should not be thought of "as a formula which can be illuminatingly introduced into any and every moral act. There exists, so far as I know, no formula of the latter kind."[14]

MORAL VISION AND *PHRONESIS*

Above I note that *phronesis* can be understood as a kind of "insight." It has qualities of vision that enable the one who possesses this virtue to *see through* (that is, see with, not see behind) the variability of human goods. For Murdoch, overcoming the fantasies of consolation requires vision, and she relies on metaphors of sight throughout both the fiction and the philosophy. As such it should not be surprising that she describes fantasy as "blinding" and that which "prevents one from seeing" what is there outside of one.

In a famous statement from her 1964 essay "The Idea of Perfection," Murdoch noted, "I can only choose within a world I can *see*, in the moral sense of 'see' which implies that clear vision is a result of moral imagination and moral effort."[15] She places this view in the minds of many of her characters. For instance, in *The Book and the Brotherhood*, after Jenkin tells Gerard, "One has to think outward, onward, into the dark," Gerard replies, "Only as far as one can *see*. After that it is fantasy."[16]

It is important here to see the contrasting pictures. If a picture has held us captive in the one sense, then another picture must help liberate us or move us toward a new vision of what we might see. According to Sami Pihlstrom, "Murdoch advises us to turn our gaze from philosophical abstractions—or, in other words, from analytic metaethics—to the endless varieties of moral (forms of) life we engage in. It is here that her ... Wittgensteinianism also becomes visible."[17] Turning to a series of questions that require *phronesis* rather than *techne*, Murdoch writes,

> Our attachments tend to be selfish and strong, and the transformation of our loves from selfishness to unselfishness is sometimes hard even to conceive of. Yet is the situation really so different? Should a retarded child [sic] be kept at home or sent to an institution? Should an elderly relation who is a trouble-maker be cared for or asked to go away? Should an unhappy marriage be continued for the sake of the children? Should I leave my family in order to do political work? Should I neglect them in order to practise my art? *The love which brings the right answer is an exercise of justice and realism and really looking.* The difficulty is to keep the attention fixed upon the real situation and to prevent it from returning surreptitiously to the self with consolations of self-pity, resentment, fantasy and despair. [...] Of course virtue is good habit and dutiful action. But the background condition of such habit and such action, in human beings, is a just mode of vision and a good quality of consciousness. It is a *task* to come to see the world as it is.[18]

This seems to me to be exactly correct. And it applies to life on the farm in ways that Murdoch probably never could have imagined. Agricultural *phronesis* seeks to answer those questions about how our life on the farm enables us to flourish. What is it that we want the farm to accomplish? What kind of people do we want to become through our care and attention to the animals and the fields and crops? What kinds of animals shall we raise? For what purposes? In what ways can the farm become a place of hospitality for guests, of study and reflection for students, of joy and learning for children and grandchildren, and of useful work and meaningful leisure for all of us? The appropriate role for *techne* arises when we seek to answer *how* we might best accomplish the goals and ends identified above. It is a rough ground indeed.

We seek to overcome the consoling fantasies of *techne* that prompt us to imagine that through rules, formulae, procedures, and technologies we might escape the difficulties of this mortal coil. These difficulties, this variability of human goods, does lead us to recognize the need for that attention and habit formation that pushes toward the difficult *phronesis* rather than the consoling *techne*. This is the pursuit of that Good that is the *eudaimonia* made possible by divine grace. It is Good, but it must not be confused with the Nice. "Back to the rough ground!"

CALVES

In "Ockham, Iris, and the Show Cattle," I have documented some of our difficulties breeding cattle. About four years into our cattle adventure, we finally began to have some success. In the span of ten months, three of our four heifers produced calves. And the birth of a calf is just about the most extraordinary thing that has happened to us on the farm.

The first to calf was Honey, the big red Beefmaster heifer we bought at the Houston Rodeo from a ranch about an hour south of Waco. Honey was artificially inseminated with the semen from a champion Beefmaster bull named "EMS Headliner." Earlier in my life I never imagined that the day would come when I would be looking through catalogues and discussing bull semen with my wife and teenage son. Times change and whole new worlds open up to us.

We had to choose two bulls because of some conflicting bloodlines in the two heifers we wanted to breed. We put in our order and the ranch released the semen (I'm not joking) to us. It was sent from a custom breeding service to our veterinarian for AI (artificial insemination). We bred both Honey and Scarlet this way. Scarlet had a miscarriage, but Honey's insemination was successful. We had known that Honey was getting close to delivery, and on a sun-splashed October day, Andrew discovered the beautiful, dark-red heifer that had just been born. We named her "Honeysuckle."

We acquired Boogie from our friends Mackie and Norma Jean Bounds at Swinging B Ranch in Axtell, just outside of Waco. Mackie is a past president of the Beefmaster Breeders Association. We bought our first heifer (Bessie) from him, and he and his team have helped and guided us throughout our cattle adventure. We took Boogie back to their ranch, and she was bred naturally to "Adonis," one of Mackie's prize bulls. In the days leading up to her delivery, Boogie's udder looked as if it were about to burst. We began to be concerned. We moved her into the pen by the barn where we could feed her and keep an eye on her. One crisp February morning, after I had already gone to work, Andrea discovered Boogie laboring in that pasture. Boogie did a great job, and Andrea was able to watch the entire delivery of a beautiful little bull. Boogie's official name is "Boogie Fever," and we named her bull "Stevie Wonder."

The last to calf was the Jersey heifer, Annabelle. Annabelle is a special case. We had bought her for $300 when she was about two weeks old. While all of our other cattle have been Beefmasters, Andrea had long dreamed of having a Jersey that we might keep for a milk cow. Most Jersey heifers that are for sale go for $700-$1000 or more. When Andrea found Annabelle for sale on Craigslist for only $300, it seemed too good to be true. She was in Greenville, Texas, almost three hours away, but we hitched up the trailer and went and got her.

She was a beautiful little calf. Andrea bottle-fed her for about three months before we gradually moved her to feed. Early on she had been sickly, and we took her to the vet and gave her a heavy dose of electrolytes. As she grew and became stronger, we began to have some doubts about her. Was it possible that Annabelle had been a twin? Twin calves of mixed genders often result in a sterile female, called a freemartin. Many dairy farmers simply won't take a risk on a female twin, so they sell them for cheap with dairy bull calves. Was that why she was so inexpensive? We feared this may be the case.

There was only one way to find out. (Okay, actually there were several ways of finding out. We could have the vet do an ultrasound and even run chromosome tests to see if she were a freemartin, but all of that would be more expensive.) Our friends Matt and Missy have a superb Akaushi bull that had sired several calves for other friends of ours. Akaushi is a relatively rare breed of Japanese beef cattle. Annabelle spent three months with them, and we were hopeful. Missy told us that Annabelle and the bull spent a lot of time together that first month, but after that they did not have much to do with one another. According to those calculations, Annabelle should have given birth in late June.

June came and went and there was no calf. Her udder had clearly expanded, but it was small in comparison to how Boogie's had looked before her delivery. And Jerseys have huge udders. Was something wrong? By mid-July, her udder was growing, but she still seemed a long way off. As we approached early August, we were coming to the last possible date when the bull could have bred her. And just then Andrea saw signs of hope. She began to see what looked like the movement of a calf in her belly. On the afternoon of the fourteenth, almost two months after we thought she should have given birth, we noticed that Annabelle had separated herself from the rest of the herd and was lying down. This is often a sign that labor is near. Andrea and I walked out into the large new pasture to check on her. She got up to greet us, but her udder still looked nothing like Boogie's had been.

This week was a busy one for us. Andrew, our youngest, was packing for college, and he and Andrea had been busy collecting and organizing all that he would need for his move to the dorm. Andrea and I were preparing emotionally for the empty nest. On Wednesday morning Andrea got up early to go to physical therapy for the rehab of her broken shoulder. As I went over to feed the animals, something caught my eye that didn't seem quite right in the pasture. Was that a deer? No. There was a new member of the herd in the pasture. And he was walking and staying close to Mama.

I sent Andrea a picture but then realized that she wouldn't be able to see it during therapy. Little did I know that Andrea had spent the morning telling the physical therapist about our anticipation of this new calf. When she discovered the pictures on her phone, she went running back inside to show her therapist the good news.

The first few hours in a calf's life are crucial. He's got to get a good dose of colostrum, and he needs to start nursing as soon as possible. Most ranchers have stories about first delivery heifers who wander off in the pasture away from their calves. I knew that it was essential that I move Mama and baby from the larger seventeen-acre pasture to the small "nursery" pen attached to the barn. The two pastures are separated by two sets of fences on either side of a gravel driveway that leads back to our house. I will have to move Mama and baby through two gates and across the driveway.

Moving the baby is easy. I pick him up and carry him to the other pen. But now Mama is upset. She is mooing and pacing back and forth along the fence line. I try to herd her toward the gate, but the other five cows have arrived and are now all crowding at the gate. When she can't get through, she retreats along the fence line. I've no option but to open the gate and try and drive Mama toward the other pen. When I open the gate, the other five cows make a dash for freedom. They rush through the opening and scatter down the driveway and around the (now dry) pond. Now that they are out of the way, I can lead Mama Annabelle through both gates and to her baby. I secure them in the nursery, and call Andrew on the cell to come help with the other cows. He gets a bucket of feed, and they all happily follow him back to where they belong.

Over the next few hours, the baby begins to nurse and settle in. On Thursday morning we move Andrew, our fifth and youngest child, to college and the dorm. Mama and baby are waiting for us when we return. We only thought we were empty-nesting.

E. B. WHITE'S ADVENTURES IN CONTENTMENT

George Orwell famously wrote, "I worshiped Kipling at 13, loathed him at 17, enjoyed him at 20, despised him at 25, and now again rather admire him."[1] Something similar could be said of my own relationship to Elwyn Brooks ("E. B.") White. My mother read and reread *Charlotte's Web* and *Stuart Little* to me before I could read them by myself. They were some of my favorite books. Charlotte and Templeton kept me from ever really despising (or fearing) spiders and rats. I have almost always been an early riser, and while I can't be certain of the reason, I think Mr. Arable's words to Avery just might have always been in the back of my mind: "Fern was up at daylight, trying to rid the world of injustice. As a result, she now has a pig. [...] It just shows what can happen if a person gets out of bed promptly."[2]

By the time I reached adolescence, I thought his books were only for children, and I was fairly certain that rats couldn't read and spiders couldn't spell. As a college undergraduate, I could now afford to be condescending and sentimental to Charlotte, but I didn't know that White wrote anything else. I was surprised to discover that the same person wrote (rewrote, that is) *The Elements of Style*, which some of my professors revered more than the Bible. With sophomoric brilliance, I declared it to be hopelessly old-fashioned and parochial. As a graduate student, I learned to refer to the book as "Strunk and White" and conceded that it had a place in "formal" writing. As a professor, I now keep a copy near my computer at all times (even if I don't always follow the hallowed suggestions).

But it wasn't until we had the farm that I really began to read White seriously or realized the depth and breadth of this beloved author. Here was a man who loved literature and culture, who made his money in the city but whose heart was on the farm. There is a refreshing honesty and humility about White's writings on farming. In one place he writes, "I have been fooling around this place for a couple of years, but nobody calls my activity agriculture. I simply like to play with animals. Nobody knows this better than I do—although my neighbors know it well enough and on the whole have been tolerant and sympathetic."[3] Sounds pretty familiar.

From the summer of 1938 to the winter of 1943, White wrote a monthly column for *Harper's* under the title "One Man's Meat." The collected edition of these columns has been in print ever since. In most of them, White reflects on experiences from his Maine farm or in the interchange between his life in New York City and his life in Maine. He touches on many aspects of farm life, and most of them are as true today as they were in the forties. In one of my favorites, he reflects on "the miracle of the by-products." He writes:

> Plane a board, the shavings accumulate around your toes ready to be chucked into the stove to kindle your fires (to warm your toes so that you can plane a board). Draw some milk from a creature to relieve her fullness, the milk goes to the little pig to relieve his emptiness. Drain some oil from a crankcase, and you smear it on the roosts to control the mites. The worm fattens on the apple, the young goose fattens on the wormy fruit, the man fattens on the young goose, the worm awaits the man. Clean up the barnyard, the pulverized dung from the sheep goes to improve the lawn (before a rain in autumn); mow the lawn next spring, the clippings go to the compost pile, with a few thrown to the baby chickens on the way; spread the compost on the garden and in the fall the original dung, after many vicissitudes, returns to the sheep in the form of an old squash. From the fireplace, at the end of a November afternoon, the ashes are carried to the feet of the lilac bush, guaranteeing the excellence of a June morning. (January 1943)[4]

Not everyone appreciated these columns. He quotes a friend who said to him, "I trust that you will spare the reading public your little adventures in contentment."[5] But this is wrong on so many levels. White's "adventures in contentment" are fabulous. They are wonderful to read in and of themselves. Like reading Boswell on Dr. Johnson or Montaigne's *Essays*, you can open to any page, and you are virtually guaranteed to find something insightful, witty, or provocative. More importantly, however, it is also the case that our culture is one desperately in need of learning how to be content. Our culture is one of reckless convenience and unrestrained consumption that is rarely, if ever, content. White's reflections are an antidote to this cultural malaise.

Moreover, it's interesting that his writings would be described as "adventures in contentment" because, for the most part, White himself was not a model of contentment. He was melancholy, frequently given to anxiety and depression, and spent much of his adult life worrying about his health. But the contentment that he did have is often found in his powers of observation and appreciation. He is content because he learned to see what was there, and having seen it, realized its true value.

One of the things that I like about the *Harper's* columns is that they were written during a time of war. The news from the front is frequently a part of these columns, almost like the radio playing softly in the background, from which he would have received so much of this news. And he is forthright about whether his reflections on collecting syrup or attending to lambs or caring for chicks around a brooder stove are worthwhile amidst all the carnage and tragedy occurring over there. He's not always sure of the answer.

I think we find ourselves in much the same position. There are wars and rumors of war on every hand. There are urgent problems that must be solved immediately, if not sooner. And there is outrage and indignation. Outrage and indignation are two of the things that our culture does best. In such a world, it might seem hard to justify the time and the money spent raising a small flock of hens or growing a variety of heirloom tomatoes or just watching pigs dig around among the roots of some great tree. In each of these cases, the task before us is to cultivate the habit of attention and to turn that attention to the faithful stewardship of those blessings I am so easily inclined to forget. For these "adventures in contentment" in the midst of an ever-present crisis, Elwyn Brooks White is a fine friend.

GUSSIE, LLOYD, AND MOCHA

Long before we bought our own farm we had some experience with donkeys. Andrea's parents created a beautiful farm in Plantersville, Texas, and her father, Ron, still keeps several donkeys there. (In truth, the "H5 Farm" in Plantersville was where our farm adventure actually began. Andrea grew up planting trees and working there on weekends, and when I entered the family, it was always a mainstay of our family get-togethers. We still gather there regularly for holidays and special events.) Ron has had numerous animals there over the years: goats, longhorn cattle, miniature horses, and all manner of ducks and geese. But his donkeys have always been our favorite. From the time our children were old enough to venture into the pens, they wanted to see the donkeys and feed them their beloved vanilla wafers.

So when our dear friends Betty and Buster needed to find a home for "Gussie," one of the miniature show donkeys they raised, we were excited to take her in. Donkeys are a great asset to a farm. Even miniature donkeys make for wonderful livestock guardians. The coyotes don't realize that the animal making that loud braying noise only stands a little higher than three feet. But her diminutive size is no impediment to her protecting the farm. Donkeys equate most dogs with predators, and they can trap the coyote inside all four of their legs and trample them to death. I've seen Gussie chase a coyote all the way off of our farm.

But she's more than a guardian. Gussie has always been one of the most popular animals on our farm. She has an easy, affectionate temperament, and children of all ages can come and pet her all over. She also ranks pretty high on the adorable chart. When our son

Samuel was trying to think of a creative way to ask a girl to the prom, he somehow managed to use Gussie to make the pitch. She said yes.

For the first few years, Gussie would hang out with the sheep or with the cattle, but she never had a companion of her own. Betty had told Andrea that we might want to think about breeding her, and for a number of months we looked for a suitable miniature jack (male donkey). Through friends at church, we finally met Michael and Jessie. They had recently relocated to Texas, and they had a beautiful miniature jack named "Lloyd." And they lived on a small farm right here in Crawford. Would they mind if Lloyd took a "vacation," and came to be with Gussie for a couple of months? They thought it was a great idea, and we went and got Lloyd.

I've mentioned in other reflections that Andrea runs an Airbnb ("The Cottage at Benedict Farms") out of the small farmhouse that we renovated when we first bought the farm. Lloyd and Gussie were in the pasture closest to the Cottage. They quickly became attached to one another, and from what we were told by some of the guests, the visiting children may have gotten more of an education about the birds, bees, and donkeys than we intended.

Lloyd stayed for a couple of months, apparently did his job, and then went home. One day a couple of months later, Andrea got a call from Jessie asking if perhaps we would like for Lloyd to come back for a return visit? It seems that they had some young goats that Lloyd was not getting along with. (She had walked out one day and found Lloyd holding a baby goat in his mouth.) Perhaps he should have another vacation.

Gussie was thrilled to have Lloyd back. They would nuzzle each other, and they never strayed too far from each other's side. And they were both still very good with guests. In time Lloyd's visit became permanent. We were glad to welcome him here for good. Though a good bit larger than Gussie, Lloyd is just as gentle and affectionate (we no longer have any goats) and (we hope) even more intimidating to the coyotes. And while we still lose chickens from time to time, I haven't seen a coyote on the farm since Lloyd arrived.

But there was another piece of the puzzle missing. We needed an animal that grandchildren could ride. Andrea had sat our granddaughter Clara on Gussie's back for pictures from time to time, but Gussie couldn't actually be ridden. We needed a pony. Andrea found a beautiful dark brown miniature horse at a farm just outside of Waco. "Mocha" was six years old and accustomed to both saddles and grandchildren.

After that it wasn't long before Clara, age two, was riding Mocha in the front pasture of the farm. Clara's feet didn't quite reach the stirrups, and I was holding the lead line, but she was riding all by herself. When her little brother, Jack, was born and we went to see him in Nashville, Clara reminded us, "I want to ride Mocha!" That's music to a grandparent's ears.

Mocha brought another advantage to the farm. When we bought her, she was already bred, and it wasn't too many months before she gave birth to a beautiful little colt we named Wendell. I don't think Mr. Berry will mind his new namesake, though I might imagine his response: "At least they didn't name a jackass after me." Hopefully Gussie will deliver in the coming year, and then if we can get Lloyd interested in Mocha, we might just hit the jackpot . . . a mini-mule!

IN DEFENSE OF WATCHING GRASS GROW

Some would have us believe that watching grass grow is the epitome of boredom. That is not true. Any cattleman worth his salt will tell you that to be in the cattle business is really to be in the grass business. Without good grass there is no future for cattle or any other livestock. It's all about the grass.

Is there any task that has consumed us more on a regular basis than tending to the grasses of our pastures and fields? I don't think so. Before we began our farm adventure, I had no idea how important the grass really was or how much of our time and energy would be spent on it. It's not that I took grass for granted. I knew that good grass took a lot of work. In Texas we mow the grass nine or ten months of the year—often from early March through December. I can recall many occasions of trying to get the grass (and the leaves) mowed before family and friends came over for Christmas parties. And in the summer, we fight the heat by constantly moving sprinklers all around the yard—and summer lasts a very long time. (There is an old joke that Texas has four seasons: almost summer, summer, still summer, and January.) In suburbia keeping the grass green, the yard neat, and the sidewalks edged was primarily a matter of pride and about being a good neighbor.

But in the country, good grass becomes a matter of life and death. If the livestock can't graze on grass, you're done. It's just that simple.

In Wendell Berry's *Remembering*, Andy Catlett remarks that one of his father's favorite sights was cattle grazing on good grass.[1] When I first read this, I did not understand the significance of the statement. I thought that the sentiment here was a nostalgic one. This too is false. Cows grazing on good grass is a present good that makes future goods possible. It's life and health. And it is hard to come by and not to be taken for granted.

The only thing better than livestock grazing on good grass is livestock *not* grazing on good grass. This is one of the points of Psalm 23 that many of us learned at our mother's knee. "He maketh me lie down in green pastures" (Ps 23:2). A green pasture is a lush, full pasture, and a sheep or a cow that is lying down in this green pasture is one that is full and content. In this moment the heifer or the ewe is not grazing because she has all that she needs or wants. It exemplifies the words of the psalmist, "I shall not want."

I think that there are many parallels in all aspects of our lives to this need for the life and health of good pastures, but for the moment, I'll stick to grass. How do you keep the grass growing at a rate that will provide sufficient forage for the livestock and yet not be abandoned and overcome with weeds or grazed down to nothing? It can be very difficult, especially when we go through long periods of drought.

At the most basic level, it means that we must be attentive to the care and maintenance of the grasses. It means regularly reseeding and, at the appropriate times, shredding and mowing to allow new growth to appear and flourish. It means "rotational grazing" of livestock—moving the stock from one field to another and allowing those unused fields to rest and recover.

But rotational grazing is far from a simple matter. To rotate the stock assumes that you have "sufficient" fields or pastures from which and to which to move your animals. What counts as sufficient will depend on how many head and how many acres you are managing. Rotational grazing requires adequate land and good fences, and fences that can keep stock contained require money, time, and expertise. (Inexperienced philosopher hobby farmers are in short supply of all of these.) Even relatively inexpensive electric fences require one to make a number of decisions based on what kind of livestock will be contained, how far the lines will run, and how far the lines will be from a power source. And what kind of power source? If the electric fence is close to a barn, you might be able to hook up an AC fence charger. Lines that are farther away will probably require either a solar charged or other type of rechargeable battery.

Moreover, your fences must be laid out in a way that ensures that every field or pasture will have a water source. You might get lucky and have a pond or tank in some fields, but in all likelihood, you're going to be running new water lines, or at the very least hoses, to water troughs in your various fields. All of this amounts to a great deal of work.

Farmers and ranchers who practice rotational grazing disagree about how it is best done. Some would have you move the livestock from one pasture to another only after the livestock have grazed an entire field *almost* completely. If you allow the stock to graze your entire property, they will wander throughout it, eating all of the good stuff and leaving only the bad weeds. On this account, after they have eaten everything, you can move them to another field that has been resting and growing thick grass.

Joel Salatin famously (or infamously) practices a different kind of rotational grazing. At his "Polyface Farm" they move the stock to a different pasture every day to enhance the diversity of vegetation growing in each field, or as Salatin calls it, "the salad bar."[2] The "lunatic farmer" says, "Never let them take a second bite." On Salatin's family farm, they attempt to replicate what happens in nature by putting poultry on the fields to follow the cattle. The chickens and turkeys scratch through the cow manure looking for bugs and grubs, breaking up the manure and working it into the soil for greater overall soil (and livestock) health. This sounds so simple, but it is not.

Did I mention that this business of watching grass grow is rather labor-intensive? It sure ain't boring. And when we have failed to attend to our grasses or when the Texas summer heat has simply burned up what little grass was there to begin with, we find ourselves buying ever more expensive hay and praying for rain.

"In the country, there ain't nothing to do but watch the grass grow."

And it's a beautiful sight.

ORCHARDS

We have a small collection of fruit and nut trees on our farm. One day at work, I made a passing reference to our "orchards," and a colleague teased me about my use of the word. "Moore, I hardly think 'orchard' is the right description of your trees." He might have been right. "Orchard" is probably a little misleading, and "our orchards" is downright pretentious. However, several sites on the internet, that great storehouse of knowledge, claim that "five trees" is the minimum number of trees required for an orchard. If that's the case, we make the cut. We have eleven mature, bearing pecan trees, and we've planted eighteen more grafted pecan saplings. We also have seven assorted varieties of bearing peach trees, two bearing olive trees, and a plum and a chestnut tree. And we have plans for lots more.

I'm happy to claim "orchard status" via internet confirmation, but I doubt there is a fixed number that makes an orchard. This is probably a version of the sorites paradox, better known as the fallacy of the "heap." (If you have a heap of sand, and you start reducing it one grain of sand at a time, when will it no longer be a "heap"? A fixed number doesn't make a heap.) But the number is not really that important. The presence of an orchard on a farm is a marker of more important matters. Orchards, perhaps more than any other feature of the farm, are a sign of the long-term planning, care, and attention required by a good farmer, and a mature orchard is evidence of many years of just that attention and care.

Every year we experiment with new varieties of vegetables in the gardens. Whether it's a new kind of tomato, a different sort of squash, or the latest variety of green beans that the feedstore is recommending, we'll try almost anything. Some of these work out and become new favorites. Lots of them wither and die in the Texas heat. Next year we'll try something else. The stakes are pretty low.

Trees, however, are a different matter. First, they are much more expensive, and most of us cannot afford to experiment with varieties that don't have a good track record in this region. Even then we don't always know what's going to work in our soil. Second, planting any kind of tree, but especially fruit trees, requires lots of patience. In most cases it will be several years before the young sapling will begin to bear its fruit. Trees also require a lot more work. Depending on its age, size, and variety, it might require a very large hole (in ground that might have a lot of rock), a mulch bed, or a berm and stakes to protect it from the wind, or even some fencing wire to protect it from livestock. In the early years, it will require a lot (but not too much) watering. There are two obstacles here. First, sometimes the trees are a long way from the water source, and you will have to haul water to them (over and over again). Second, finding that mean of how much water is enough is not always easy. We have planted many, many trees that have not made it.

And once it does begin to bear, its care is only magnified. It will need to be pruned carefully since some branches will only bear fruit once. If the tree is a dwarf variety or small in stature, the weight of the fruit might seriously damage the tree, and its limbs may need to be supported by additional stakes and wires. In *A Place on Earth*, Wendell Berry describes Mat Feltner's loving care of his orchard: "Feeling the limb on which the ladder is propped spring against his weight as he moves, Mat prunes the tree. He likes this work—the look of his hands moving and choosing, correcting, among the tangle of the branches. The orchard is one of the works of his life."[1]

Pruning is not easy work. It's easy to get carried away and start cutting limbs that ought not to be cut. That's why it's better to use a handsaw or clippers for pruning fruit trees. The chain saw makes it all too easy, and when the work gets easy it starts to go fast. This is, of course, a blessing and a curse. The blessing part is easy: a chain saw can cut a lot of wood in a short period of time. But the curse follows in kind: because it's so easy I start cutting everything I can reach. Sometimes I have found myself doing ridiculously stupid things—like standing under a limb I'm about to cut, or worse, standing on my tiptoes and, with the running chain saw held aloft in one hand, straining to reach just one more branch. Benjamin would remind me of the old Bill Engvall song "Here's Your Sign."

Since the handsaw requires hard work, it slows you down. It's always best to capitalize on one's essential laziness, and it's much better to find yourself wondering which limbs MUST I cut rather than which ones CAN I cut. You can always go back and cut more, but when you've sawed off the limb that shouldn't be pruned, there is no going back.

That's why the presence of that mature orchard is such a sign of the longevity of the farm and the temperance and patience of its farmers. It just takes years and years of care to grow an orchard. In one of his essays, Czeslaw Milosz speaks of returning to the river valley of his birth and childhood in Lithuania. All of the small villages and farms, along with their barns, gardens, and orchards, had been cleared and replaced by huge fields for the Soviet collectivist agricultural endeavor. Milosz writes, "Among the many definitions of Communism, perhaps one would be the most apt: enemy of orchards. For the disappearance of villages and the remodeling of the terrain necessitated cutting down the orchards once surrounding every house and hut." Milosz continues:

> Orchards under Communism had no chance, but in all fairness let us concede that they are antique by their very nature. Only the passion of a gardener can delight in growing a great variety of trees, each producing a small crop of fruit whose taste pleases the gardener himself and a few connoisseurs. Market laws favor a few species that are easy to preserve and correspond to basic standards. In the orchards planted by my great-grandfather and renewed by his successors, I knew the kinds of apples and pears whose very names pronounced by me later sounded exotic.[2]

I want to have the passion of that gardener that Milosz describes. We have certainly had the delight. Our best success with fruit trees has been with peaches. We have peach trees of several different varieties and ages (on account of the ones that have died and been replaced). This means that we have a relatively long season during which our peaches are ripening on the trees. We enjoy the fresh peaches over ice cream on hot summer evenings, and Andrea makes a lot of peach preserves. Eight pounds of peaches will give us eighteen to twenty jars of preserves, and because the trees are ripening at different times, we're able to repeat this process two or three times over the summer.

And our pecans are even more productive. The individual trees will offer a pecan crop in different years, but we almost always get enough to give away bags and bags of both fresh shelled pecans and candied pecans at Christmas. We get the vast majority of the pecans "rough shelled" for us at a processing plant outside of town. This means we still have a lot of work to do to pick out excess pieces of pecan shell before they can be bagged or cooked. And by the end of the year, we often have an extra fifty to sixty pounds that never got shelled or candied. We sell those directly to the feedstore that posts the daily buy rate outside by the road.

So far we only have two olive trees, but we've been impressed with the small batch of olives they produce. For a number of years, I have been taking students to study in Greece and Turkey, and I marvel at the thousands of acres of olive groves we see there. Among my more ridiculous delusions is my dream of producing enough olives for us to have both a nice collection of olives in brine and enough to press for our own olive oil. It will obviously take many more trees than we currently have (but I know where we're going to plant them).

There is another reward to growing a small orchard of fruit trees that only comes a couple of times a year. Imagine a hot summer day when you are mowing the grasses around the peach trees just as their fruit is becoming ripe. As you make several passes by and under the trees, you begin to inspect their offerings. "This tree is little bit behind that one." "Those peaches on that upper limb are looking particularly good." And then, perhaps drenched in sweat, with more than a little dust and grass clippings on your clothes, you make your selection, reach up, and pluck a prize peach straight from the tree as you drive by.

"Do I dare?" Yes, you do.

CITY OF SOWS

In Book II of Plato's *Republic*, there is a famous distinction between "the city of sows" and the "luxurious city." I've taught this material for many years, and while there are many fascinating and provocative aspects to this part of the dialogue, the notion that the first city is a "city of sows" is not one of them. I've always taken it for granted that Glaucon (Plato's much older brother and one of Socrates' chief conversation partners in this dialogue) was right about the first city. And then we got pigs.

The distinction arises because Glaucon is not content with the first attempt that Socrates and Adeimantus (another of Glaucon's and Plato's brothers) make at constructing a theoretical city that will help them explore the notion of justice. The brothers Glaucon and Adeimantus have been demanding that Socrates give them a better version of the argument for why justice is its own reward. If Socrates is going to praise justice—for its own sake and not merely for its consequences—then he will have to show why a just man who is believed to be unjust will be happier than an unjust man who not only goes unpunished but also has a reputation for being just. Plato is of course setting up a picture of the historical Socrates himself, a just man who was hated, convicted, and executed for alleged injustice while demonstrating what a life of true flourishing and happiness should be.

Socrates tells Glaucon and Adeimantus that he cannot give them the persuasive argument that they want, but he also cannot *not* attempt to do so. He suggests that they seek to understand first what justice in the city looks like (justice written in large letters) and then see if it's the same in the individual (justice written in small letters). The first attempt is made by Adeimantus and Socrates, who describe an ideal city in which all of the citizens recognize that they have different talents and they can all work together to meet the bodily needs of the citizens. Allan Bloom describes the first city this way: "It is an easy place: there is no scarcity, and justice takes care of itself. Men join together because they are incomplete, because they cannot provide for their needs themselves. Their intention is not to have more than others but to have enough for themselves.... In such a city there is no need for men to be governed."[1] Glaucon is not satisfied with the austere city that his brother and Socrates have imagined, and thus he describes it as a city of sows. Socrates and Glaucon then imagine a second city, "the luxurious city," and how justice and injustice might naturally grow and develop there.

No one to my knowledge questions Glaucon's label of the first city as a city of sows. But of course it's not at all what a city of sows would look like. In the city of sows, it's "root, hog, or die." Scarcity is not an issue with pigs. It does not matter how much food they have. They want what the other pigs have. You can pour feed into two different and separated troughs, and the pigs will fight over the food in the one before moving to the other (and then fight over it there as well). Justice only "takes care of itself" in the most ruthless of ways.

Likewise, there is a certain luxury about the way pigs make their nests and beds. As noted in "Not So Humble, But Near to the Ground," the pig is one of the very few farm animals who will not foul his or her sleeping quarters. In fact, they will go to extraordinary lengths to dig, root, and cover themselves in mud or straw when they make their beds. To the untrained eye, the pigpen looks like a muddy mess, but on closer inspection, one can see that by almost any standard, pigs are extraordinarily careful about their beds and their pens. One might even say that they luxuriate about their beds. (This illustrates how cruel the modern industrial "farrowing pens" are for pigs. These stainless-steel crates are designed so that the sow will not crush her newborn pigs. But she cannot even turn around in these crates, let alone root and make the nest that her instincts demand. And of course she must now defecate in the one place where she both lives and sleeps. In theory her excrement should fall through the cracks in the floor, but the theories don't always work out.)

Back to the nest. What are we to make of all this? Is it perhaps the case that Plato simply knew very little about pigs and thus he has Glaucon dismiss the first city as a city of sows because he imagines that the grunts and squeals are devoid of any intelligence or higher goods? Stanley Rosen emphasized that it is merely the *cuisine* of such a city that is fit for sows.[2] Leo Strauss famously read the Platonic dialogues as possessing a hidden "esoteric" meaning that often undermined the explicit "exoteric" meaning of the text. Is there a hidden meaning that we are supposed to see here? I don't think so.

I think Plato just did not know pigs. In fact, it's not just Plato. Philosophers in general seem rarely to have thought very seriously about pigs. When they have, they've given in to the same kinds of gross generalizations about pigs that are common throughout our culture. Contrary to how we philosophers like to understand ourselves, we've accepted a conclusion without examining the evidence, namely that the basest forms of existence are those befitting pigs. The most damning thing Thomas Carlyle could say about utilitarianism was to equate it with being a "philosophy fit for swine" (on the premise that utilitarianism was only about the maximizing of pleasure over pain). In response John Stuart Mill famously retorted, "It is better to be a human being dissatisfied than a pig satisfied; better to be Socrates dissatisfied than a fool satisfied. And if the fool, or pig, is of a different opinion, it is because they only know their own side of the question."[3]

Pigs are not fools. Any farmer will tell you that they are just about the smartest animal on the farm. Perhaps this is why Orwell puts the pigs in charge of the revolution in *Animal Farm*. And if Plato or Carlyle or Mill or any other great lover of wisdom thought that a base, merely sensual existence is best described by the example of the pig, it's simply more evidence of how these great thinkers failed to understand one of the fundamental insights of philosophy: Never assume that appearance and reality are the same. Always recognize that "how things are" might be different from "how they seem to be."

This insight will of course be one of Socrates' primary points of emphasis in the pages leading up to and following from the famous "allegory of the cave" that appears at the beginning of Book VII. "It's the nature of the real lover of learning to struggle toward what is, not to remain with any of the many things that are believed to be" (*Rep.* 490b). What distinguishes the lovers of wisdom from both the Sophists and the Lovers of Opinion is that they understand the difference between appearance and reality.

The allegory of the cave illustrates both the highs and lows that follow. The prisoners trapped in the cave first assume that the shadows on the wall are real. After they have been unchained and turned around they will see the puppets and the fire and know the "truth"—or at least part of the truth. But then, mistaking the fire for the sun, they will be tempted to settle down by the fire, content with a little bit of knowledge that they confuse with the wisdom they claim to want and to love.

A city of sows can teach us a lot, but we won't learn anything until we recognize what we do not know. And this, of course, Socrates knew.

FARMING WITH THE PHILOSOPHERS

WORK, LEISURE, WONDER, AND GRATITUDE

"True philosophy consists in learning anew to look at the world."

—Maurice Merleau-Ponty

We have a fair number of guests who come out to the farm. Student groups, church groups, friends of our children who are looking for a quiet place to recoup and regroup, be it only for a few hours or for a few days. This morning Samuel and two fraternity brothers showed up to sit on the porch, play guitars, and generally get ready for their coming final exams. As mentioned earlier, Andrea runs an Airbnb in the Cottage that was the original farmhouse on the property. Some of these guests (like some of my students) have never been on a farm before. They have never fed animals, collected eggs, watched a cow being milked, or seen a newborn calf up close. For these people, especially the children, the farm is a place of wonder and excitement.

Sometimes we will have guests at the farm who grew up on a farm or perhaps they remember a family farm from their childhood. Sometimes there is wonder and joy in these people as well, but more often than not, they remember farm life very differently. As our dear friend Suzanne said to us, "We did it. You can have it." Collecting eggs meant trying not to get pecked by broody hens or attacked by an aggressive rooster. (My father had an experience with a rooster during his Mississippi childhood that forever colored his view of raising chickens.) Milking the cow meant cold mornings in a dark barn in which you were lucky if you only got kicked or swatted by her tail. It meant long days and few "city luxuries." It was hard work.

I have thought a lot about the relation between work and wonder, toil and leisure. The cheap and easy answer is that "wonder" and "leisure" are the luxuries of pretend farmers like us who are not depending on the rain to pay the bills. There is bound to be some truth in that. We do not live and die at the mercy of the prices at the grain elevator or the sale barn. One does not negotiate with the elevator. You get what you get, and far too often it is not nearly enough. But that easy answer does not account for the superlative experience of wonder that we regularly see from some of the hardest-working farmers and ranchers we know. When one of our first heifers was calving, Andrea called our ranching friend Lyndon with a question. After answering her question, he commented on the joy of watching a calf born. "I've seen it a million times. And every time it's a miracle."

Where does the wonder come from, and where does it go? I think the wonder comes from not only being perplexed by something but by realizing that a particular kind of perplexity is not merely a problem to be solved but a world to be explored. It is the recognition that there is something here that is greater than I am and beyond my immediate ability to grasp or control it. Aristotle famously argues in the *Metaphysics* that philosophy begins in wonder.

Where does the wonder go? Well, there are two senses of "go"— where does it lead and why does it leave? I'm more interested in the former, so I'll take the latter first. Wonder "leaves" because the toils and cares of life and work deceive us into believing that we do not have time for it or that it does not matter. There is too much to do to stop and reflect. Wonder sounds a lot like daydreaming, and daydreaming does not get the work done. And why reflect upon the work anyway? Often it is uninteresting and those to whom I am accountable do not care about what I think about it and do not want me wasting my (or their) time reflecting on the meaning of the universe. This all too common view reinforces the notions that work is drudgery and the worker is merely a functionary.

There is much more to be said about this. The loss of wonder in our work (whether at home, school, office, factory, or field) leads to a boredom and an experience of meaninglessness that is all too common in the modern world. The modern world is often described an age of *ennui*. This French word is usually translated as "boredom." In truth, it is actually that weariness of the world that comes from our boredom with it. It is a world that is devoid of wonder, and in a world like ours in which technology and automation play ever greater roles, it becomes all the more common.

"Success" and wealth are no barriers to *ennui*. Walker Percy memorably described "the anomie of the late twentieth century, [and] the cold phlegm of Wednesday afternoons" this way: the successful businessman who comes home from work to the suburbs every day at 5:30 "and there is the grass growing and the little family looking not quite at him but just past the side of his head, and there's Cronkite on the tube and the smell of pot roast in the living room, and inside the house and outside in the pretty exurb has settled the noxious particles and the sadness of the old dying Western world, and [him] thinking: 'Jesus, is this it? Listening to Cronkite and the grass growing?'"[1] That's *ennui*.

I have often thought of the farm as the cure for *ennui*. It is impossible (for me) to get bored on the farm or to believe that our actions or inaction does not matter to all of the beings with whom we reside here on the farm and in the networks of relationships that spin out from it. The barnyard and the gardens and the fields and the ponds are endlessly fascinating and liberating for me—to say nothing of all of the people one encounters in this world. Of course I recognize that this experience is not universal. Many people experience rural and farm life as being trapped rather than liberated. Rural America is in a crisis for many reasons, but perhaps the greatest is the terrible exodus that has occurred as people have left the relative isolation of the rural areas for the "expanded opportunities" of the cities. The rural loss of jobs, inadequate education and social services, and limited technological access have sent untold millions into the cities and left our rural communities even worse off than they were before. Ironically, however, it is the opportunistic cities that are most responsible for the epidemic of *ennui*.

Wonder does not pay the bills. But merely paying the bills will not satisfy the soul either. We have to find a way to rediscover the wonder that can come from a life in which our work is transformed through a renewed recognition of its significance, or at least the possibility of its significance. But that renewal rarely (if ever) comes from the work itself. It comes from understanding the work we do as part of something greater, something of worth and significance. It comes from a transformed understanding of leisure.

LEISURE

When most of us hear the word "leisure," we think of relaxation or "downtime." Whether pursuing our favorite hobbies, slumping on the couch watching a movie or a game, or venturing forth at our favorite restaurants, stores, or vacation spots, we have a tendency to think of leisure as a hedonistic (in a good way) activity (or inactivity, as the case may be). But the best sense of leisure is far more significant than merely the pursuit of our various delights and pleasures.

Few books have been as important to me in my adult life as Josef Pieper's *Leisure: The Basis of Culture*. Written in a postwar Germany that was divided between a capitalist West, which called for a Weberian ethic of total work to rebuild culture and society, and a communist East, which understood itself as celebrating and defending the dignity and worth of the Worker, Pieper offers a magnificent rejoinder to the world of total work proposed on both sides of the Wall. Leisure is not mere recreation or release. It is not inactivity as opposed to active work. It is not merely the restoration of one's energy and power so that one can return to work. All of these things have useful benefits, but they are not true leisure.

Leisure in Greek is *skole*, from which the Latin *scola* and the English "school" are derived. (My students find it uproariously funny that "school" and "leisure" have the same root.) But it is true. The years that are set aside for education were not traditionally for "job training"; they were instruction in those arts that are truly liberating, hence the "liberal arts." True leisure requires not only the time and space to attend to what is most important, it also requires the personal inclination to stop and attend to those matters. Our true freedom consists of our ability to cultivate a disposition that inclines us to recognize and distinguish between the various matters that matter.

For Pieper, following both Aristotle and St. Thomas Aquinas, authentic leisure is an attitude of mind and a condition of the soul that recognizes that our true liberation cannot be willed in the way of total work and acquisition. Flourishing will not be achieved through the will to power. Flourishing will only be possible through the habit of attention. Murdoch confirms the same: "Will cannot run very far ahead of knowledge, and attention is our daily bread."[2] And attention is only possible when we stop, look, and listen to the world around us.

Pieper puts it this way: "Leisure is a form of silence, of that silence which is the prerequisite of the apprehension of reality. [...] Leisure is a receptive attitude of mind, a contemplative attitude, and it is not only the occasion but also the capacity for steeping oneself in the whole of creation." For Pieper our connection to the "wholeness" of creation is most important. Our work so often pushes us to think of ourselves and our lives only as functionaries. On this view we are never more than cogs in the wheel. Certainly there is some truth here. We all play small roles that are parts of larger goods. But Pieper recognizes that when we become entirely absorbed in our limited functions, we lose our capacity (and our desire) to see the world as a whole. We are not meant for fragmentation. We are meant for Wholeness. "Because Wholeness is what man strives for, the power to achieve leisure is one of the fundamental powers of the human soul. [...] Only in genuine leisure does a 'gate to freedom' open."[3] For Pieper the truest expression of leisure is ultimately found in worship.

Given the multitude of tasks that we have, it is easy to see how we are not only pushed and pulled in all that has to be done, but how we also begin to equate "leisure" with simply a "break" from the work. This both blinds us to the reality of our situation and proposes a false solution to our dilemma. We should not be surprised when we discover that this view of leisure consistently fails: either the "break" does not satisfy (and thus we try for ever greater or more expansive such breaks) or we refuse to "break" at all because it distracts us from what is important, that is, the "work."

The goal is not a "balance" between work and leisure. The goal is a life lived in which genuine, contemplative leisure guides and informs both the ends and means of our work. It means that we have to think differently about knowledge itself. Pieper is clear that a sense of *affirmation* is at the heart of any true sense of leisure. And this affirmation takes me out of myself and my petty obsessions and into a larger, more expansive recognition and affirmation of what is.

Murdoch sees something quite similar. In "The Sovereignty of Good over Other Concepts," she writes,

> I am looking out of my window in an anxious and resentful state of mind, oblivious of my surroundings, brooding perhaps on some damage done to my prestige. Then suddenly I observe a hovering kestrel. In a moment everything is altered. The brooding self with its hurt vanity has disappeared. There is nothing now but kestrel. And when I return to thinking of the other matter it seems less important. And of course this is something which we may also do deliberately: give attention to nature in order to clear our minds of selfish care. [...] It is so patently a good thing to take delight in flowers and animals that people who bring home potted plants and watch kestrels might even be surprised at the notion that these things have anything to do with virtue.[4]

For Murdoch, this observation is about the centrality of beauty to our lives. I want to go in a slightly different direction, without denying Murdoch's insight. Yes, true Beauty does delight the soul, and since it also reveals the Good and the True, it should never be underestimated. But the crucial issue for me here is the connection between wonder, wholeness, and gratitude (which of course is a reflection of the perfection that is Beauty). This takes me back to the earlier question of where the wonder *leads*. Wonder might lead to superficial curiosity or it can lead to a more sustained engagement that is often (but inadequately and misleadingly in my view) called "studiousness." When wonder leads to sustained study and engagement, then the best result is gratitude. This requires some unpacking.

CURIOSITY AND STUDIOUSNESS

The common distinction between *curiositas* and *studiositas* is a good place to start. Curiosity is the superficial desire to know a little bit about something or find some helpful information or acquire some knowledge only for the purpose of owning it or using it for a discrete, particular task. And while curiosity famously "killed the cat," we should recognize that it is a natural intellectual appetite, and one that we all employ regularly about things that we need to know but about which we do not recognize any transformative significance. "Studiousness," on the other hand, is an attitude that recognizes that what is under examination requires my sustained attention; it defies my ability or inclination merely to use it or to acquire it. It is more than a desire for "information." I like the way Paul Griffiths puts it.

> Where curiosity wants possession, studiousness seeks participation. They also differ in the kinds of knowledge they seek. Curiosity is concerned with novelty: curious people want to know what they do not yet know, ideally, what no one yet knows. Studious people seek knowledge with awareness that novelty is not what counts, and is indeed finally impossible because anything that can be known by any one of us is already known to God and has been given to us as unmerited gift.[5]

The difficulty with the term "studiousness" is all its academic connotations. It makes it look like this attitude or disposition only applies to those enrolled in school, or to geeks or nerds or academics. But there are more ways to understand "school" than just "formal schooling." If we think of "school" in the sense of true leisure explained above, then whenever we are carefully attending to the matters that matter, whether we are at home or in the office or in the barn, we are "being schooled" in the most important sense. And we must be appropriately "studious" if we are to succeed in this larger, more comprehensive "school of life."

In this reading true wonder and studiousness are intimately connected. Curiosity stands on its own. Above I said that "wonder might lead to superficial curiosity." Now we can see that true wonder does not lead only to curiosity. True wonder leads to the kind of sustained engagement I have described as studiousness. For example, if we are "wondering" how best to keep the sow from crushing her farrow of pigs, then we might buy or construct a "farrowing pen" that will allow her little pigs to nurse but will keep them safe from her accidentally rolling over on them and crushing them to death. Unfortunately this pen will also defy all of her natural instincts to make a nest and to root around in the care of her pigs. But since the reason we have pigs is to breed them and thus to sell and process her offspring, it makes sense to do everything we can to maximize our investment in the pigs while minimizing not only the risk to her pigs but also the time and effort we have to spend on them.

However, if we are "wondering" what is the best and most comprehensive way that we can raise both the sow and her pigs, we may not be content to answer this question according to industry standards and practices. If we want the sow to be able to follow her instincts, to root around and make her bed in such a way that allows her "pigness" to flourish (to use Joel Salatin's word), then we are going to have to do things differently. We are going to have to construct a different sort of pen. We are going to have to figure out how to separate this sow from the others when farrowing time comes. We still want to sell and process her offspring, but now we are reflecting on this question in light of a series of larger, more comprehensive questions. It will entail learning more about pigs, watching their interactions, talking with people who know.

Moreover, any plans we make for the pigs fall within the larger questions of what do we want the farm to become, and what, if any, role pigs should play on such a farm? Now we have turned our attention toward something that requires our sustained reflection and participation. And there is probably not just one answer to these questions. There are numerous questions that we will discuss, reflect upon, and attempt to answer in both "theory" and "practice." What are we physically capable of accomplishing? What can we afford? How will these decisions affect the rest of the farm? The individual questions that have to be answered are not fragmented from the whole; they reflect the whole, and their answers return to contribute to and transform the whole. Apart from their reference to the whole, they cannot even be approached.

This is a point that Wendell Berry makes over and over again in both his fictional and nonfictional works. Our technological society has mistakenly believed that it could extract "information" from that knowledge that is first learned through love and participation. As a society we have come to believe that discrete facts about the material world, abstracted from the whole and from the context in which they grew and were cultivated, can be *put to use* in ways to satisfy our own curiosity. Even worse, the "value" of the knowledge is judged entirely on the basis of its "usefulness." In Berry's words, "Wonder has been replaced by a research agenda."[6]

Jack Baker and Jeffrey Bilbro do an excellent job of showing how the distinction between curiosity and studiousness plays out in Berry's fiction. In *Jayber Crow*, Troy is "curious" and Mattie is "studious." Troy is so keen to increase production that he does not know the real value of the land, and in his reckless pursuit of efficiency, he destroys the land by seeking to see how much he can force it to produce. He does not love the farm or farming. In fact, he resents it, and in his resentment, he uses it up. His wife, Mattie, on the other hand, loves the farm, and she preserves the old forest of hardwood trees that her father had set aside (the "nest egg") and that her husband wants to liquidate to pay his debts on the broken tractors and equipment that litter his fields. Likewise, in *Hannah Coulter*, when Hannah's son Caleb becomes a professor of agriculture, his love for the farm is replaced by the curiosity of abstraction. Baker and Bilbro conclude that Caleb "pursued agricultural knowledge not to be a better farmer or to take better care of his farm but to publish papers in the 'Unknown Tongue' and get tenure."[7]

Returning to Griffiths, notice how he turns toward the notion of knowledge as *gift*. He continues, "The deepest contrast between curiosity and studiousness has to do with the kind of world [...] each inhabits. The curious inhabit a world of objects, which can be sequestered and possessed; the studious inhabit a world of gifts, given things, which can be known by participation, but which, because of their very natures can never be possessed."[8]

GRATITUDE

The notion of "gift" is an important one in contemporary philosophy. (Leave it to the philosophers to take something beautiful like "gift" and turn *that* into a research project.) Whether we are thinking about the notion of a fact or a principle that is "given"—accepted as true without justification—or our experience of how the phenomena of the world present themselves, there are aspects of this question that are important to both "analytic" and "continental" philosophers (even if they mean very different things by their usage of the words).

There is an easy connection between what we know about the world and the extent to which we recognize that our knowledge has the character of gift or "givenness." In German this relationship is explicit in the very words. The common everyday phrase "Es gibt" is usually translated "there is." Translated literally, however, the words mean "it gives." Martin Heidegger was particularly interested in this phrase "Es gibt." (The notion of givenness was exceedingly important to his teacher Edmund Husserl, and much of the modern tradition of phenomenology is dedicated to understanding what is entailed in this givenness.)

Heidegger believed that the phrase "Es gibt" naturally inclines us to ask about who or what is doing the giving. For Heidegger, "Being" is what gives, and this simple insight provides a clue to understanding the arc and development of his thought throughout his life. In his early work, he was keenly interested in understanding the nature and character of "Being"—as both a noun (the "what" of being) and a verb (the "how" of being). There is an emphasis on action and activity in this early work because he is trying to understand "*how we be.*" In his later work, he turns toward a greater interest in *receptivity*. *How we be* becomes more about how we receive the world than how we act within it.

I cannot go into the great tragedy that is the life and work of Martin Heidegger. Richard Rorty said somewhere that Heidegger was "a liar, a coward, and the greatest philosopher of the twentieth century." I think that is pretty good, even if I do not think that philosophers can be ranked like boxing champs ("the greatest?"). I do think that there are moments of insight in Heidegger that rival any in twentieth-century philosophy. But the moments of insight are coupled with a hubris and a cruelty that virtually destroys them.

Heidegger, however, is an extraordinary companion on the farm. He had a deep sympathy for and recognition of the vitality of rural life and the ways in which we human beings use tools to accomplish our many goals. He understood, theoretically at least, the necessity of human limits, and I believe that these limits play an important role in the turn toward receptivity. The central project of his great (and incomplete) 1927 book *Being and Time* is to offer an examination of what it means to be human. Or, rather, to understand the particularly human way of being human in a particular place and at a particular time. Right at the heart of this understanding of not only who and what kind of thing we are, but also what it means to be at all, is the recognition of limits. In that book Heidegger appeals to the "quiet power of the possible" (a phrase he will return to in the "Letter on Humanism").

Wendell Berry and Wes Jackson are fond of quoting Alexander Pope's 1731 admonition that one should "consult the genius of the place." In their usage this is an encouragement to understand what a piece of land can and cannot do. We "consult the genius of the place" when we slowly learn what sorts of vegetation will flourish here or how the rainwater runs off there or how it responds to the changes of the seasons. When we consult the genius of the place, we do not rush in and impose our will on the land. The Dust Bowl tragedy of the 1930s, when American farmers cultivated millions of acres of wheat on prairies that were ill suited to sustain them, is exhibit A in the *failure* to "consult the genius of the place."

One of my favorite Wendell Berry essays is "People, Land, and Community" in *Standing by Words*. Its primary subject is the danger of turning knowledge into information, but along the way he addresses numerous subjects, including marriage, Dante on Odysseus, harmony, memory, and the implications of introducing a tractor to the farm. "Information," like the tractor, is dangerous because by reducing wisdom to bite-sized "facts," it deceives us by speeding things up and shortening our work. Berry concludes by recognizing that "time is a bringer of gifts."[9] (Heidegger, who thought that temporality reveals being, would have agreed.) When I read Jackson's exhortation to "consult the genius of the place" or Berry's recognition that "time is a bringer of gifts," I think about reflecting on the quiet power of what is possible in this place.

In a series of lectures from the early 1950s, Heidegger engaged in a lengthy reflection on the nature of thinking and on what we mean by describing an experience as "thought-provoking." One of his conclusions appealed to both an Old English etymology and the German Pietist confession that "Denken ist Danken"—thinking is thanking. This is not the place to go into all of the details about this, but the two phrases "quiet power of the possible" and "thinking is thanking" (one from the early Heidegger and one from the later) are regular reminders to me about the role of contemplation in the practice of cottage farming.

I recognize of course that my reading and use of Heidegger here is at odds with much of the standard reading and interpretation of his work. (Though I do find some consolation in being in a long line of Christians who have happily "misread" his work.) Heidegger began his work as a Catholic philosopher, but any conventional appeal to the notion of God dropped out early on. It would be wrong to imply that Heidegger's reading of "thinking as thanking" is somehow analogous to St. Paul's exhortation to the Thessalonians that one should "in everything give thanks" (1 Thess 5:18). But that's pretty much the goal.

The best sorts of thinking, the kinds of thinking that result in studiousness and not just curiosity, result in gratitude. Real thinking, or rather, the most important kinds of thinking, inclines one to recognize one's indebtedness and thus one's need for gratitude. We cannot think very hard or very long without recognizing the need for gratitude. Perhaps I should rephrase that. Of course one can think without being thankful—it is happening all the time—but one cannot rightfully think without being thankful. And one is always thankful for MORE than one currently understands. Real thinking inaugurates wonder and gratitude.

When we recognize that knowledge is gift, it changes everything.

The larger problem is that in the modern world all knowledge (and thus the world itself) becomes reduced to a commodity to be acquired and used. Participation and gratitude drop out as nostalgic sentimentality. It should not surprise us that the "information age" is an age in which technology is seen as the key to all of our problems. Of course, one of the ironies here is that technology is often proposed as the means for achieving those very "labor-saving" devices that will enhance our "leisure." Only now this "leisure" is meant as merely a break from the drudgery of labor and to release us from the work that might have taught us gratitude.

Hopefully nothing in this book suggests that farming is "leisurely" (in the vulgar sense) or that the work of farming is not hard, difficult, back-breaking work. Most of the vignettes and essays are designed precisely to illustrate how demanding and difficult farm work is in both the physical and mental challenges it presents to farmers. And these days the stresses of the financial commitments farmers face are often simply overwhelming. As the old adage goes, "Hard work ain't easy." But farming at its best is also part and parcel with the "love of wisdom." And those of us who have been fortunate enough to see this world through the eyes of those who have fallen in love with wisdom (true "philosophers," including those farmers, ranchers, and gardeners who embody this love) are reminded that the gifts of wonder and gratitude are intrinsic to this difficult and rewarding life.

So what's the point of all of this? Some reader might ask, "Does this joker actually think people will rediscover the wonder in farming if they just read Heidegger?" No. The point is definitely not Heidegger. In the most expansive sense, the point is not even about farming. Making the necessary substitutions for clothing and equipment, most of what I've said in this book is equally applicable to work in the office, factory, classroom, or home. It's just that on a farm it is harder to miss the reality, the real-ness, of life because it isn't hidden behind a computer screen or sealed off under the manufactured comfort of fluorescent lighting and air conditioning.

This book began with the story of the death and burning of Daisy the goat. That episode happened rather early in our farming adventure. The little boy that cried as we loaded a dead goat into a wheelbarrow is now in college. Since then we have seen numerous deaths on the farm: the long line of hens taken by all sorts of predators, including our neighbors' dogs and one magnificent red-tailed hawk who treats our farm as if it were his own favorite restaurant; the poor witless guinea fowl who fearlessly fight snakes during the day but foolishly roost in the trees at night, only to have their heads bitten off by owls; the stillborn piglets; and the beautiful little lambs, some of which are taken by coyotes and some of which simply cannot survive the challenges of life. And this is to say nothing of our own role in "harvesting" or "processing" those portions of our livestock. Life and death go together—which we certainly see off of the farm as well. No, the point is definitely not Heidegger.

The point is the gift. It is receptivity to gift. And in the case of the farm, it is the recognition that a life that is lived not only in close proximity to nature but also through the midst of good, strenuous work is most conducive to the cultivation of that authentic leisure that is necessary for significance and thus for true liberation. The practical arts and the liberating arts are mutually reinforcing. Learning to farm or to do any meaningful work is, as Griffiths reminds us, learning to inhabit "a world of gifts, given things, which can be known by participation, but which, because of their very natures can never be possessed."[10]

APPENDIX

IRIS MURDOCH'S VEXED RELATIONSHIP WITH CHRISTIAN FAITH

[The following essay is not about farming or agricultural matters. It is included here because Murdoch features prominently in a number of the other essays in the book. Since Murdoch explicitly denied the existence of God, some readers might find my appropriation of her work in a theological context to be unfair, deceptive, or self-contradictory. In this essay I offer an explanation of my reading of her work on this important subject. An earlier version of this essay was initially read at the International Iris Murdoch Symposium at the University of Chichester in the Autumn of 2017.]

Some years ago I read a description of Iris Murdoch as "the most religious atheist of the twentieth century." I do not remember who said it, but it certainly rings true. Murdoch was quite explicit about her disbelief in the existence of God (as conceived by orthodox Christian faith). At times she expresses this disbelief in a casual, offhanded way that suggests that it is simply obvious that traditional belief in God is untenable in the modern age. At other moments she speaks with great solemnity about the difficulties of living in a world without God. On any account, she rarely makes sustained, philosophical arguments against God's existence, probably because she believes them to be not only unsuccessful but also unnecessary. When she does engage that tradition of argument, especially St. Anselm's Ontological Argument, her conclusions are deeply counterintuitive to the spirit of her age. And yet the question of religious belief and practice recurs again and again throughout her philosophical works, her fiction, and in interviews and remembrances by friends and colleagues. At moments she speaks longingly about religious belief and even commends certain religious practices like prayer and the sacraments, among others. Whatever religious belief is, she will neither affirm it nor let it go.

Religious skepticism and disbelief in the modern age are usually consequences of what Max Weber identified as the disenchantment of the world, but this is not the case for Murdoch. Her world remains extraordinarily enchanted. The source of her disbelief is not disenchantment. What are we to make of this apparent incongruity? The canonical, though I think problematic, answer that even a novice reader of Murdoch's work knows to give is that Murdoch substitutes a Platonic notion of the Good for the idea of a personal God. Her words ring out: "Good represents the reality of which God is the dream."[1] This Good is said to perform most of the functions of God but without the embarrassing mythological anthropomorphisms or the consoling fantasies of notions like redemptive suffering. I think this answer is problematic—both for her (given her own commitments) and for her readers.

I want to reflect on three questions: First, does the "Good for God" swap work for Murdoch? Is it a persuasive solution to the age-old dilemma? Second, what is the relation for her between disenchantment and disbelief? Is enchanted disbelief an option? And third, what are we to make of her very specific, perhaps even self-indulgent, appropriation of certain religious sensibilities without God? I think that trying to answer these questions might help us make some progress on understanding Murdoch's "vexed relationship to Christian faith."

I do not mind showing my hand here. I am not at all sure of my answers to these questions. I think the crucial problem is the role of consolation, especially the construction of those consoling fantasies that Murdoch does such a persuasive job of otherwise exposing in both her philosophical and fictional works. I think each of these three questions is bedeviled by consolation. And I'm inclined to think Murdoch knew this too—which is partly why she can never turn the page on religious belief.

GOOD FOR GOD?

The standard answer is that the Platonic notion of the Good functions in the place of the traditional conception of God. We get versions of this argument in "The Sovereignty of Good over Other Concepts," in "On God and Good," in "The Fire and the Sun," and at various moments in the Gifford Lectures, available in published form as *Metaphysics as a Guide to Morals*. We also see this on display in numerous characters, both sympathetic and unsympathetic, in the novels: Marcus Fisher, Tallis Browne, Rupert Foster, Guy Openshaw, Anne Cavidge, Jenkin Riderhood, Uncle Theo, and Stuart Cuno, among others, make various versions of this argument or simply assume that some version of it is the case.

In its defense, the argument has many advantages: Good is undefinable, ineffable, incorruptible, mysterious, magnetic, alleged to be generative of love, not a thing among things, "resists collapse into the selfish empirical consciousness," and perhaps most importantly, is not subject to mythological and personal anthropomorphisms or falsifiable historical claims.[2] When we turn to evaluate these claims, it is clear (as Murdoch also knew) that everything said about the Good above, apart from the mythology and the history, can certainly be, and has been, predicated of the Christian God. No orthodox Christian theologian I have ever read believes that God is a thing among other things.

It seems doubtful that it is the mythological character of the Christian narrative that Murdoch objects to because she is not opposed to the myths and the metaphors. In her 1962 review of Elias Canetti's *Crowds and Power*, she noted that "the mythical is not something 'extra;' we live in myth and symbol all the time."[3] On many occasions she praises the nature and role of myth. When speaking of Plato's dialogue the *Meno* in her Giffords, she says, "The happy almost playful *ironic light* (invisible to many readers, such as Buber and Heidegger) in which these beautiful world-changing constructions are set out, also and constantly gives us the clue to understanding them."[4] The same is true with respect to the biblical stories. When Jonathan Miller seeks to get her to disavow them in an interview, she replies, "But we live by stories. I mean, we interpret stories; we are always using metaphors and stories. I don't myself see any problem there at all. I still myself use the Christian mythology. I am moved by it and I see its religious significance and the way in which ordinary life is given a radiance." She continues that some people will say, "'But it's not true.' Well, now, about Christianity, in a sense, all right. *I* don't think it's true that there is a heaven, or God, and so on, but this seems to me not at all to touch the sacredness and the numinousness and the truth-bearing properties of the Christian mythology."[5]

It might, however, be the claim to history that is the problem. In *Metaphysics as a Guide to Morals*, she says, "If one does want to believe in what is 'deep' in the form of the old God then let this belief be kept mysterious and separated and pure, and not mixed up with dubious history, or indeed with any history."[6] But this has always been a problem for the Christian faith, as Murdoch clearly knew. We see it addressed in the Gospel of Luke when Luke is at pains to put down historical markers about when such and such happened (Jesus is born when "Cyrenius was governor of Syria" [Luke 2:2]) and in the Creed when it is confessed that Christ is "crucified under Pontius Pilate." Here too, Murdoch is at pains not to disavow too much. When Miller asks her about religious belief despite the fact that the "story is completely untrue," she responds, "Well, yes, but you are using this terminology which I don't like—'The story is completely untrue.' The story of Christ's life is perhaps partly true. It's a story about a man. The notion that he rose from the dead is perhaps untrue."[7]

But it is not the Good's distance from history or mythology that is the signal feature that elevates the Good over God for Murdoch. The most important feature of the Good is that it "has nothing to do with purpose, indeed it excludes the idea of purpose. [...] The only genuine way to be good is to be good 'for nothing' in the midst of a scene where every 'natural' thing, including one's own mind, is subject to chance, that is, to necessity. That 'for nothing' is indeed the experienced correlate of the invisibility or non-representable blankness of the idea of Good itself."[8]

From the Christian point of view this elevation of the "for nothingness" seems very odd. It is reminiscent of the sort of "conceptual idolatry" that figures like Jean-Luc Marion have been so successful at exposing. This is reminiscent of the so-called "God of the philosophers" who argued that God *must be* a maximally perfect yet simple instantiation of all of the great-making properties (omnipotent, omniscient, omnipresent, etc.). But why? What is it about these criteria that makes one so confident that *they must be ascribed to God*, or else this being cannot rightly be considered God? It seems to me that Murdoch is giving in to a similar temptation with respect to the Good when she claims that it *must be* "for nothing." Why?

The reason that "for nothingness" is so important is that she wants to safeguard Goodness from utilitarian corruption. From any who would think that Goodness could be pursued as a penultimate Good. She knows that many religious people have loved God, not out of love for Godself, but out of desire for reward or out of fear of punishment. "Not for nothing" but for something. And that desire for something turns out to be the real God that these people have worshipped—be it power, pleasure, honor, safety, or prosperity.

I think that Murdoch's "for nothingness" plays precisely the same role that the Liberal assumption of the necessity of neutrality for moral recommendations played in the "current view" of morality that she critiqued in her early essays in moral philosophy. The vigorous rejection of attaching morality to the substance of the world was ultimately based, not on the notion that there are no metaphysical entities nor on the narrowly logical argument that one cannot derive 'ought' from 'is,' but on the comprehensive competing moral view that one must not impose one's own moral views on others. Murdoch is explicit; it is the broad, ambiguous moral view that does the heavy lifting for the behavioristic moral philosophy of the 1950s: "I think that much of the impetus of the argument against naturalism comes from its connection with, and its tendency to safeguard, a Liberal evaluation."[9] Murdoch beautifully exposed the truth that this was precisely what had been done. About the modern moral philosopher, Murdoch notes that he has not produced "a model of any morality whatsoever. It is a model of his own morality."[10]

Turning back to the question of the whether the "for nothing" is a necessary characteristic of the Good, we should listen to

Murdoch's own recommendation: "We should, I think, resist the temptation to unify the picture by trying to establish, guided by our own conception of the ethical in general, what these concepts *must be*."[11] By the same token, I am not persuaded that Murdoch is justified in prescribing antecedent conditions for what her own Good *must be*, or to be more precise, that the criteria she settles on are necessary or persuasive.

But perhaps more to the point, Murdoch's Good turns out to be not "for nothing" either. Despite all of her protestations, and the beautiful, elegant passages that so many of us love to quote, the reality of the Good, "the magnetic centre towards which love naturally moves,"[12] is a source of immense consolation, and it is the essential element that undermines the behavioristic moral philosophy she did battle with most of her days. She recognizes this, and she attempts to avoid this problem in "The Sovereignty of Good" when she speaks of true morality as a "sort of unesoteric mysticism, having its source in an austere and unconsoled love of the Good."[13] But is it unconsoled? She explicates this Good through an examination of Plato's metaphor of the sun. "It gives light and energy and enables us to know the truth. In its light we see the things of the world in their true relationships."[14] This is an ultimate consolation.

But of course there is more. Two of the most important characteristics of the Good are that it is generative of Love and that it discloses reality to us. Through Love we are able "to see and to respond to the real world," and this produces authentic humility, "selfless respect for reality and one of the most difficult and central of all virtues."[15] And none of this would be possible without the Good. Love's "existence is the unmistakable sign that we are spiritual creatures, attracted by excellence and made for the Good. It is a reflection of the warmth and light of the sun."[16] Not a bad consolation, that.

But all of this just illustrates that Murdoch's Good, as she herself notes, is as praiseworthy of the virtues and as susceptible to many of the difficulties as the notion of God was and is. In "On God and Good," she imagines an interlocutor who makes just this objection to her. One might say, "To speak of Good in this portentous manner is simply to speak of the old concept of God in a thin disguise. But at least 'God' could play a real consoling and encouraging role. It makes sense to speak of loving God, a person, but very little sense to speak of loving Good, a concept. 'Good' even as a fiction is not likely to inspire, or even be comprehensible to, more than a small number of mystically minded people who, being reluctant to surrender 'God,' fake up 'Good' in his image, so as to preserve some kind of hope." She responds to these objections with the observation, "I am often more than half persuaded to think in these terms myself."[17]

But she does not abandon her view of the Good. She believed that when one looks seriously at human things, one sees that "there is more than this." "The 'there is more than this,' if it is not to be corrupted by some sort of quasi-theological finality, must remain a very tiny spark of insight, [. . .] it seems to me that the spark is real, and that great art is evidence of its reality." And finally, "The image of the Good as a transcendent magnetic centre seems to me the least corruptible and most realistic picture for us to use in our reflections upon the moral life."[18]

Whether one finds this move persuasive will largely depend on how capacious one's views of both God and Good are and whether and to what extent both God and the Good are susceptible to human manipulation and corruption. The twentieth century surely brought ample demonstration of both. More to the point, we should ask just how nervous we are about "corruption by some sort of quasi-theological finality." And that takes us to our question about disenchantment. If Murdoch does not reject an enchanted universe, why does she insist on a disenchanted God in the form of an enchanted Good?

ENCHANTED WORLD, DISENCHANTED GOD?

So what does Murdoch mean by "quasi-theological finality," and why is she wary of it? I think that for Murdoch, theological discourse tended to close down debate rather than open it up for further exploration and conversation. If God is a moral fact, then perhaps there is nothing else to be said. The Good, on the contrary, seems to be open to endless conversation and reflection. More importantly, the vast majority of her community (including her husband and most of her colleagues) had long ceased to think of religious or theological discourse as offering a persuasive account of the nature and character of the world. Such discourse was at its best in ceremonial and aesthetic forms and occasionally useful for understanding emotive states of affairs. Of course, she had numerous counterexamples at her disposal: Peter Geach and Elizabeth Anscombe, for starters, as well as people like Eric (E. L.) Mascall, Basil Mitchell, Austin Farrer, and Michael Ramsey[19] and others. But the significance of these paled in comparison to the impact of friends, colleagues, and lovers like her husband, John Bayley, or friends such as A. J. Ayer, Philippa Foot, Elias Canetti, and many more. A. N. Wilson reports that Murdoch told him over lunch that John Bayley "got rid of God" for her.[20]

There is nothing novel here. We live in an age of secularity, and, particularly in academic and intellectual circles, belief in God is today sometimes more surprising than disbelief. The short answer (let us call it the efficient cause) to the question, "Why does Murdoch not believe in God?" is because she lived in an era and in a context within which it had largely become obsolete. This is the disenchantment of the world. As Max Weber formulates the notion, this disenchantment of the world is one of the signal features of the modern age. A sense of the sacred, promoted and supported by magic, superstition, myth, religion, and tradition, gives way to an instrumental rationality in which wonder and awe are replaced by scientific investigation. As Weber puts it in the famous address "Science as a Vocation," "There are no mysterious incalculable forces that come into play, but rather that one can, in principle, master all things by calculation."[21]

The really interesting thing to me, however, is not that Murdoch "got rid of God" but that she is so hesitant and skeptical of the general modern project of disenchantment. "Calculation" is almost always viewed disparagingly. It is explicitly stated in the philosophical essays and writ large in her fiction. And even with respect to religious belief, she seemed to break with her community in many ways that are celebratory and appreciative of religious belief and practice. In 1956 she could say that "there are no philosophical proofs of the existence of God, but it is not senseless to believe in God."[22] This would not have been a universally accepted proposition. Religious belief and theology positively haunts *Metaphysics as a Guide to Morals*. There are, no doubt, many reasons why it took her almost a decade to edit and rewrite her Gifford Lectures (including the onset of the disease that would ultimately silence her), but I have to believe that she spent much of this time struggling with what to do with a God she could neither believe in nor abandon. Near the beginning of *Metaphysics*, she writes, "We yearn for the transcendent, for God, for something divine and good and pure, but in picturing the transcendent we transform it into idols which we then realise to be contingent particulars, just things among others here and now." And by the end of it, she will say, "God does not and cannot exist. But what led us to conceive of him does exist and is *constantly* experienced and pictured."[23]

One of the charitable explanations for her counterculturalism is that Murdoch is keenly aware of the need for humility and caution when it comes to speaking about the possibility (or the impossibility) of the transcendent. She lauds the importance of "reflection, reverence, respect."[24]

Murdoch is most explicit about her unique and personal appropriation of certain elements of religious belief—mostly Christian, but also Buddhist and Hindu in places—in a 1988 interview with Jonathan Miller, quoted above, which is reprinted in Gillian Dooley's excellent collection of interviews and conversations, *From a Tiny Corner in the House of Fiction*. In the interview Murdoch is quite clear about her denial of belief in the existence of a personal God, but she is also insistent on the validity of personal appropriation. She believes, "People should be able to realize that they can have Christian religion without literal dogma, that they can have a religious dimension in their lives without having to subscribe to beliefs that now seem to them impossible."[25]

But of course there is nothing new about these beliefs seeming to be impossible. They were always problematic. They always seemed impossible. St. Paul says, "We preach Christ crucified—a stumbling block to the Jews and foolishness to the Greeks" (1 Cor 1:23). That Murdoch believes that she can have those parts of religious experience that are helpful and inoffensive without the challenging and unsettling parts sounds very much like the construction of the kinds of consoling fantasies that Murdoch vigorously attacks throughout both her philosophical and fictional works. It is the difference between the comfortable Nice and the difficult Good.

And I think Murdoch recognized this. When Miller asked her if her participation in religious services was anything more than "touristic enthusiasm," she replied, "Good heavens, yes. I do. I mean, I don't actually take communion, I don't receive the sacrament. Something inhibits me from doing this—I'm not quite sure what. But I certainly attend Christian services. Religious experience is something we should be having all the time in fact [...] religion is something that fills the whole of one's life. It's to do with every moment in one's life."[26]

Those of us who are religious wholeheartedly agree, and it is a great consolation for us—as it clearly was for Murdoch. Whether its consoling character invalidates it is another question.

NOTES

Preface
1   Wendell Berry, *Imagination in Place* (Berkeley: Counterpoint, 2011), 13.

Homecoming and the Future of Higher Education
1   Wes Jackson, *Becoming Native to This Place* (Berkeley: Counterpoint, 1996), 3.
2   David W. Orr, *Earth in Mind: On Education, Environment, and the Human Prospect* (Washington: Island Press, 2004), 119.
3   Orr, *Earth in Mind*, 119–20.
4   Orr, *Earth in Mind*, 120. Aldo Leopold, *The River of the Mother of God and Other Essays by Aldo Leopold*, ed. Susan L. Flader and J. Baird Callicott (Madison: University of Wisconsin Press, 1991), 302 (cited in Orr, *Earth in Mind*, 120).
5   Joel Salatin, *The Marvelous Pigness of Pigs: Respecting and Caring for All God's Creation* (New York: Faith Words, 2016).
6   Orr, *Earth in Mind*, 120.
7   An earlier version of this essay was published by *Plough Magazine* under the title "Why Higher Education Is Failing to Prepare Students for the Future (and What Schools Can Learn from Our Agrarian Roots)." See www.plough.com/en/topics/community/education/why-higher-education-is-failing. Used with permission.

The Fallacy of Acquisition
1   Boethius, *The Consolation of Philosophy*, trans. Victor Watts (New York: Penguin, 1999), 35 (2 P 5).
2   Wendell Berry, *The Art of Loading Brush* (Berkeley: Counterpoint, 2017), 193.

To a Hare, from a Louse

1   Robert Burns, *Poems*, ed. Henry Meikle and William Beattie (London: Penguin Books, 1946).

Farmers, Christians, and Intellectuals

1   Robert Bellah et al., *Habits of the Heart: Individualism and Commitment in American Life* (New York: HarperCollins, 1985).
2   Robert Jenson, "Triune Grace," *Dialog* 41, no. 4 (2002): 285–93.
3   Josef Pieper, *A Brief Reader on the Virtues of the Human Heart*, trans. Paul C. Duggan (San Francisco: Ignatius Press, 1991), 37–38.
4   Iris Murdoch, *The Nice and the Good* (New York: Penguin, 1968), 187; and Murdoch, "On 'God' and 'Good,'" in *Existentialists and Mystics*, ed. Peter Conradi (New York: Penguin, 1999), 342.
5   Iris Murdoch, "The Fire and the Sun," in *Existentialists and Mystics*, 459.
6   G. K. Chesterton, *Orthodoxy* (1908; San Francisco: Ignatius Press, 1995), 25.
7   Iris Murdoch, "The Idea of Perfection," in *Existentialists and Mystics*, 331.
8   Murdoch, "The Idea of Perfection," and "The Sovereignty of Good over Other Concepts," in *Existentialists and Mystics*, 333, 376–77.
9   Iris Murdoch, "Literature and Philosophy: A Conversation with Bryan Magee," in *Existentialists and Mystics*, 26.
10 Murdoch, "The Sovereignty of Good over Other Concepts," in *Existentialists and Mystics*, 373.
11 Wendell Berry, *A Place on Earth* (Washington, D.C.: Counterpoint, 1983), 206.
12 Josef Pieper, *Faith, Hope, Love*, trans. M. F. McCarthy (San Francisco: Ignatius Press, 1997), 99–100.
13 Pieper, *Faith, Hope, Love*, 100.
14 Pieper, *Faith, Hope, Love*, 113, 117.
15 All quotations in this paragraph are from Pieper, *Faith, Hope, Love*, 118–19 (block quotation from 118).
16 Pieper, *Faith, Hope, Love*, 124–25, 126.
17 Pieper, *Faith, Hope, Love*, 129.
18 Pieper, *Happiness and Contemplation*, trans. Richard and Clara Winston (South Bend, Ind.: St. Augustine's Press, 1998), 84–85.
19 Pieper, *Happiness and Contemplation*, 105–6.

DEAD LAMBS

1 Gene Logsdon, *Holy Shit: Managing Manure to Save Mankind* (White River Junction, Vermont: Chelsea Green Publishing, 2010), 82.

ALEXANDER MCCALL SMITH

1 Alexander McCall Smith, *The Forgotten Affairs of Youth* (New York: Pantheon, 2011), 158.

2 Alexander McCall Smith, *The Minor Adjustment Beauty Salon* (New York: Pantheon, 2013), 103.

OCKHAM, IRIS, AND THE SHOW CATTLE

1 Murdoch, "The Sovereignty of Good over Other Concepts," in *Existentialists and Mystics*, 371, 375.

WENDELL, GENE, AND JOEL

1 Kevin Lowe, in *Baptized with the Soil: Christian Agrarians and the Crusade for Rural America* (New York: Oxford University Press, 2016), does a superb job of documenting and narrating the Protestant half of this history.

2 Wendell Berry, *This Day: Collected and New Sabbath Poems* (Berkeley: Counterpoint, 2013), xxi.

3 Wendell Berry, foreword to Gene Logsdon, *Living at Nature's Pace*, rev. ed. (White River Junction, Vt.: Chelsea Green Publishing, 2008), p. vi.

4 Logsdon, *Holy Shit*, 54.

5 Todd S. Purdum, "High Priest of the Pasture," *New York Times*, May 1, 2005.

6 Michael Pollan, *The Omnivore's Dilemma: A Natural History of Four Meals* (New York: Penguin, 2006), 133.

7 Pollan, *Omnivore's Dilemma*, 124.

8 Wendell Berry, foreword to Logsdon, *Living at Nature's Pace*, vi. Wendell Berry, foreword to Gene Logsdon, *Letter to a Young Farmer: How to Live Richly without Wealth on the New Garden Farm* (White River Junction, Vt.: Chelsea Green Publishing, 2017), ix.

9 Berry, *Art of Loading Brush*, 180.

10 Berry, *Art of Loading Brush*, 185.

11 Gene Logsdon, "Wendell Berry: Agrarian Artist," in *Wendell Berry: Life and Work*, ed. Jason Peters (Lexington: University of Kentucky Press, 2007), 243.

12 Gene Logsdon, *The Contrary Farmer's Invitation to Gardening* (White River Junction, Vt.: Chelsea Green Publishing, 1997), 108, 121.

13 Preface to Joel Salatin's *The Sheer Ecstasy of Being a Lunatic Farmer* (Polyface, 2010).

14 Jeff Bilbro, "When Did Wendell Berry Start Talking Like a Christian?" *Christianity and Literature* 68, no. 2 (2018).

15 Wendell Berry, *Place on Earth*, 104.

16 Gene Logsdon, *You Can Go Home Again* (Bloomington: Indiana University Press, 1998). Gene Logsdon, *Gene Everlasting: A Contrary Farmer's Thoughts on Living Forever* (White River Junction, Vt.: Chelsea Green Publishing, 2014), 39.

17 Logsdon, *You Can Go Home Again*, 37.

18 Logsdon, *Gene Everlasting*, 154.

19 Logsdon, *Gene Everlasting*, 40.

20 Salatin, *Marvelous Pigness of Pigs*, 57.

Not So Humble, but Near to the Ground

1 William Cobbett, *Rural Rides*, ed. Ian Dyck (London: Penguin Books, 2001), 347.

2 Mark Essig, *Lesser Beasts: A Snout-to-Tail History of the Humble Pig* (New York: Basic Books, 2015), 20–21.

3 P. G. Wodehouse, *Blandings Castle* (Woodstock: Overlook Press, 2001), 86.

Saving Spiders

1 Iris Murdoch, *The Philosopher's Pupil* (London: Vintage Classics, 2000), 61.

2 Iris Murdoch, *Under the Net* (London: Vintage Classics, 2002), 52.

Back to the Rough Ground

1 Boethius, 44 (2P8), my translation. Chesterton, *Orthodoxy*, 35.

2 Wittgenstein, *Philosophical Investigations*, trans. G. E. M. Anscombe (New York: Macmillan, 1953), §107.

3 Wittgenstein, *Philosophical Investigations*, §115.

4   Walker Percy, "Diagnosing the Modern Malaise," in *Signposts in a Strange Land* (New York: Farrar, Straus and Giroux, 1991), 210–11.

5   Jacques Barzun, *The Culture We Deserve* (Hanover, N.H.: Wesleyan University Press, 1989), 39.

6   Wendell Berry, *Sex, Economy, Freedom, and Community* (New York: Pantheon Books, 1993), 142.

7   Iris Murdoch, *The Sovereignty of Good* (London: Routledge, 2001), 100.

8   Murdoch, *Sovereignty of Good*, 100.

9   Murdoch, *Sovereignty of Good*, 88.

10 Murdoch, "Literature and Philosophy," in *Existentialists and Mystics*, 4.

11 Murdoch to Jean-Louis Chevalier, "Closing Debate, *Recontres avec Iris Murdoch*," cited in *From a Tiny Corner in the House of Fiction*, ed. Gillian Dooley (Columbia: University of South Carolina Press, 2003), 78.

12 John Haffenden, "John Haffenden Talks to Iris Murdoch" (London: Methuen, 1985), cited in Dooley, *From a Tiny Corner*, 125.

13 Murdoch, "The Idea of Perfection," in *Existentialists and Mystics*, 320.

14 Murdoch, "The Idea of Perfection," in *Existentialists and Mystics*, 334.

15 Murdoch, "The Idea of Perfection," in *Existentialists and Mystics*, 329.

16 Iris Murdoch, *The Book and the Brotherhood* (London: Vintage Classics, 2003), 244.

17 Sami Pihlstrom, *Pragmatic Moral Realism: A Transcendental Defense* (Amsterdam: Rodopi, 2005), 25.

18 Murdoch, "The Sovereignty of Good over Other Concepts," in *Existentialists and Mystics*, 375 (original emphasis only on "looking" and "task").

E. B. White's Adventures in Contentment

1   George Orwell, "Kipling," *New English Weekly*, January 23, 1936.

2   E. B. White, *Charlotte's Web* (New York: HarperCollins, 1980), 5.

3   E. B. White, *One Man's Meat* (Gardiner, Maine: Tilbury House Publishers, 1997), 143.
4   White, *One Man's Meat*, 275.
5   Scott Elledge, *E.B. White: A Biography* (New York: W. W. Norton, 1984), 216.

In Defense of Watching Grass Grow

1   Wendell Berry, *Remembering* (Berkeley: Counterpoint, 2008), 57.
2   Joel Salatin, *Salad Bar Beef* (Swoope, Va.: Polyface Publishing, 1996).

Orchards

1   Wendell Berry, *Place on Earth*, 163.
2   Czeslaw Milosz, "Happiness," in *To Begin Where I Am* (New York: Farrar, Straus, and Giroux, 2001), 24.

City of Sows

1   Allan Bloom, "Interpretive Essay," in *The Republic of Plato*, trans. Allan Bloom (New York: Basic Books, 1991), 344.
2   Stanley Rosen, *Plato's Republic: A Study* (New Haven: Yale University Press, 2005), 75.
3   John Stuart Mill, "Utilitarianism," in *The Basic Writings of John Stuart Mill* (New York: Modern Library, 2002), 242.

Farming with the Philosophers

1   Walker Percy, "Bourbon," in *Signposts in a Strange Land*, 103.
2   Iris Murdoch, "The Idea of Perfection," in *Existentialists and Mystics*, 335.
3   Josef Pieper, *Leisure: The Basis of Culture*, trans. Alexander Dru (San Francisco: Ignatius Press, 2009), 46–51.
4   Murdoch, "The Sovereignty of Good over Other Concepts," in *Existentialists and Mystics*, 369–70.
5   Paul J. Griffiths, *Intellectual Appetite* (Washington, D.C.: Catholic University of America Press, 2009), 22.
6   Wendell Berry, *The Way of Ignorance* (Berkeley: Counterpoint, 2005), 55.

7   Jack R. Baker and Jeffrey Bilbro, *Wendell Berry and Higher Education: Cultivating Virtues of Place* (Lexington: University of Kentucky Press, 2017), 178–79.
8   Griffiths, *Intellectual Appetite*, 22.
9   Wendell Berry, "People, Land, and Community," in *Standing by Words* (1983; Washington, D.C.: Shoemaker and Hoard, 2005), 77.
10 Griffiths, *Intellectual Appetite*, 22.

Iris Murdoch's Vexed Relationship with Christian Faith

1   Iris Murdoch, *Metaphysics as a Guide to Morals* (London: Penguin, 1992), 496.
2   Murdoch, "The Sovereignty of Good over Other Concepts," in *Existentialists and Mystics*, 376.
3   Murdoch, "Mass, Might, and Myth," in *Existentialists and Mystics*, 191.
4   Murdoch, *Metaphysics as a Guide to Morals*, 475.
5   Dooley, *From a Tiny Corner*, 214–15.
6   Murdoch, *Metaphysics as a Guide to Morals*, 472.
7   Dooley, *From a Tiny Corner*, 217.
8   Murdoch, "On 'God' and 'Good,'" in *Existentialists and Mystics*, 358.
9   Murdoch, "Vision and Choice in Morality," in *Existentialists and Mystics*, 95.
10 Murdoch, "Metaphysics and Ethics," in *Existentialists and Mystics*, 67.
11 Murdoch, "Metaphysics and Ethics," in *Existentialists and Mystics*, 75.
12 Murdoch, "The Sovereignty of Good over Other Concepts," in *Existentialists and Mystics*, 384.
13 Murdoch, "The Sovereignty of Good over Other Concepts," in *Existentialists and Mystics*, 376.
14 Murdoch, "The Sovereignty of Good over Other Concepts," in *Existentialists and Mystics*, 376.
15 Murdoch, "The Sovereignty of Good over Other Concepts," in *Existentialists and Mystics*, 376, 378.

16 Murdoch, "The Sovereignty of Good over Other Concepts," in *Existentialists and Mystics*, 384.
17 Murdoch, "On 'God' and 'Good,'" in *Existentialists and Mystics*, 358–59.
18 Murdoch, "On 'God' and 'Good,'" in *Existentialists and Mystics*, 359–61.
19 Murdoch and Mascall debated "Rational Existentialism" in the Oxford Socratic Club, March 3, 1952.
20 A. N. Wilson, *Iris Murdoch as I Knew Her* (London: Arrow Books, 2004), 145.
21 Max Weber, "Science as a Vocation," in *From Max Weber: Essays in Sociology*, ed. H. H. Gerth and C. W. Mills (New York: Oxford University Press, 1946), 139.
22 Murdoch, "Vision and Choice in Morality," in *Existentialists and Mystics*, 93.
23 Murdoch, *Metaphysics as a Guide to Morals*, 56, 508.
24 Murdoch, *Metaphysics as a Guide to Morals*, 462.
25 Dooley, *From a Tiny Corner*, 215.
26 Dooley, *From a Tiny Corner*, 215.

www.ingramcontent.com/pod-product-compliance
Lightning Source LLC
Chambersburg PA
CBHW030650230426
43665CB00011B/1034